W9-BQS-811

The Charting of an Atmospheric Environment

Michael S. Hamilton, Ph.D.

VANTAGE PRESS
New York

FIRST EDITION

Published by Vantage Press, Inc.
419 Park Ave. South, New York, NY 10016

Manufactured in the United States of America
ISBN: 0-533-14722-0

Library of Congress Catalog Card No.: 2003095742

0 9 8 7 6 5 4 3 2 1

Contents

Acknowledgments v
Photo Acknowledgments xiii

1. The Establishment of Awareness 1
2. Invisible Potentials 19
3. Discernible Behaviors 56
4. The Cohesion of Energy 69
5. Construction of an Environment 85
6. When Molecules Gather 113
7. Protection from Above 137
8. Creative Transformation of Thought 164
9. The Acquisitions of Observations 181
10. A New Ability to Perceive 201
11. The Words in the Sky 248

References 259

Acknowledgments

This work and the effort that it contains would not have been possible without certain and specific elements. Not all elements are easy to define, as would be in the presentation of text, philosophical language, and ontology. Much was required that had to be learned and could not have been removed from other scientific publications that deliberate facts of science or of cultural history. Perseverance in research and adaptation in presentations, inclusive of general background information and philosophical interpretation, are but a few of the special nuances that are often more than not required and therefore not extractable from text. Because of these invisible and nondirect involvement requirements, individuals with abilities that stand out with certain attributes among many sometimes pass into your life who bear listening to. And special energies are then dedicated to understanding these distinguished gifts. Listed at the end of this acknowledgment section are those individuals who possess a certain something that cannot be retrieved from text but rather must be learned by example through their presence and friendship. Because of that unique learning experience, gathered and applied by our friendship, these few are left until last and a short statement has been rendered on their behalf denoting their unique and special gifts that I am sure augmented this effort. I do wish to state that the pictured individuals did not directly contribute to the work in a physical sense—that is, not through writing, research, funding, typing, expositional or contextual presentation, and so forth. In addition, because of the fact that they are not specifically considered authorities in academic disciplines of physics, philosophy, meteorology, particle theory, quantum mechanics, and so forth this publication cannot in any way be considered a forum of their knowledge; they, however, because of their continued advocation of my efforts have deserved a special place. Because of that, I wish to honor each of them with a portrait.

I wish to thank the University of New Hampshire and the people at the library and its bookstore for all their valuable assistance. Thank you to Mr. Richard Sardinha for all his assistance in accomplishing a myriad of property tasks that otherwise would have delayed this publication. Thank you to the Amazon Book Company for the availability of hundreds of texts that were required to allow for this accomplishment. Thank you to Richard O'Hearn for his assistance in the software and its implementation, which enabled the text to be compiled. Furthermore, the fundamental research concepts that were employed within the text are in no small way a result of the review of many issues of *CEP—Chemical Engineering Publications.* For his most generous efforts in supplying me with monthly reference issues I would like to thank my friend and neighbor, Mr. David Colby, publisher.

Thank you to Pacific Western University for their advice on such matters. A special thanks to S. W. Childs Management Corporation and to their president, Mr. William Ellis, specifically, for their understanding in the work and effort manuscripts such as these require. Thank you to the Jitto's Company under the direction of Mr. Daniel Nadeau and Mr. Cory Hussy, their patient manager, for allowing me to observe meteorological patterns and to discuss them with my friends while tying up one of their television sets, tuned to the Weather Channel (for so . . . long). Thank you to Mrs. Cory Hussey (Nikki), for her creative adaptation of my pictorials on some of the systems covered within the textual body and cleaned up under Mendes-Hussey Graphics. I wish to thank the American Geophysical Union, the papers under Composition and Chemistry and published under the *Journal of Geophysical Research,* for the guidance and learning that they have given. Thank you to Linda White of Rye Scapes for her photographic eye and the assistance provided at site locations for the individual pictures. Furthermore, there have been many others who have given strength and influence through their friendship, and that also qualifies as support.

For this reason, I would like to thank the following friends: Jack Gobbi, Sylvia Todlowski, Richard Groves, Shirley Ferguson, Debbie Hanley, Robert Arnold III, Robert Agresti, Dan Bicknell, David and Susan Anderson, Harold Ecker, David Dawley, Robin Moulton, Gloria Blake, Bethany Marshall, Richard Cambell, John De Lucia, Richard Sardinha, Robert Macdonald, Greg Raymond, Adam Barnfather, Michael and Teresa Robinson, David Short, Robert Amero Jr., Sharon Rainey, Barry Riddle, Andrew Widden, Michael and Lisa Shaw, Joseph Buenvoyage, Mark Lamprey, Robert March, Richard "Tony" and Linda Brunetta, Linda White, Cory Hussey, Scott Hussey, Gerald Nadeau, Pat McKeel, Peter J. Aikens Sr., Peter "Petey" Aikens Jr., "Robo" Todd, Wiliam C. Giles, Marelyn Jameson, Andy Klimchuk, Joe and Mike Gambino, Bruce

Gilbert, Destiny Doane, and Leslie Montelle Cole III and his wife, Paula Swist Cole. From each I have learned much, and to each I am grateful.

A special thank you to my wife, Lynda, and our family for all their support during the long hours over several years spent on research and text. Thank you to my brother Pat and his wife, Mary, and to my sister Libby. This publication is dedicated to my parents: Mr. Christie Patterson and Cynthia Neilson Repp Hamilton Jr.

Erik Chris Aspen

In many cases, individuals who come into contact with authors normally contribute with the direct body of the text. This particular individual, however, has contributed something entirely different from textual facts. This particular contribution was in the form of strength and perseverance. Mr. Aspen was in the submarine forces attached to the U.S. Navy, in particular diesel submarines of the Korean era (WW II Fleet Class). There is an energy that was extracted from our friendship, and because of his experiences in overcoming the hardships associated with those submarine duties he had to develop an inner strength that most of us will never comprehend. I have been blessed to observe and learn from him. It is here embodied and now applied to the research of texts and the dogmatic effort required to interpret them, which has become evident well beyond philosophical discussion. Adversity, perseverance, integrity, and detail are all part of this individual's contribution.

Matthew McQuallin Morton

This individual and his friendship over the years here in Rye have been quite revealing in the observance of Mr. Morton's ability to create with his hands and the exceptional way he can perceive another meteorological point of view. He has spent a great many hours upon the sea, and that in itself says a great deal about the necessity to modify structures due to meteorological events that required an endeavor to make one's way back to home port. Weather conditions at sea are vastly different from land phenomena because of the sea state that ocean bodies will subject one to. Especially with his experiences in small craft, Mr. Morton's ability, I believe, can be directly associated with the frailty that he has experienced while subject to oceanic hostilities. With this as a background, Matthew has contributed greatly to the observations of atmospheric progression and its expectancy. He has provided important reverence for the concern to what may be forthcoming later that day or night.

James Edwin Ray

Mr. Ray has been my friend through many years and has continually demonstrated an ability to ask questions that stimulate thinking. Meteorology is a science that continually requires asking why. Mr. Ray does this with exceptional clarity and in such a way as to make the question almost profound. The answers invariably prompt additional research necessary to adequately cover a response. In many cases, a different avenue of facts must be uncovered to satisfy the deeper "why" beyond the simplistic answer. Mr. Ray has contributed much to prompt research that I would not have otherwise endeavored to examine at a specific time within a specific chapter. His training within the U.S. Marine and Army forces has enabled him to pass on significant training and knowledge in uncovering detail that will lie just below the surface but cannot be uncovered easily.

Mitchell Edward Debose

This individual is a naval instructor, chief petty officer, U.S. Navy, responsible for teaching naval history and geography within the academic posture of naval ROTC and its associated programs. We have spent many hours discussing the atmosphere in general and as a subject that would have academic meaning in the Navy as a whole. Meteorology is taught by the Navy as part of the academics an officer would require in the line of duty, especially at sea.

Chief Debose has contributed to the concepts that academic texts would need for a purpose in presenting an understanding to individuals who require such material. His friendship and the reasoning that he has given to me has been vital in my efforts to create a meteorological foundation that can be understood through a holistic and presentable methodology.

Matthew McQuallin Morton

James Edwin Ray

Mitchell Edward Debose

Erik Chris Aspen

Photo Acknowledgments

The author would like to thank the following sources for their artwork:

Abbott, Patrick L. *Natural Disasters*. Dubuque: Wm. C. Brown, 1996. **Figs: 3.2**

Ahrens, Donal C. *Meteorology Today*. Los Angeles: West Publishing Co., 1994. **Figs: 3.3, 5.8, 5.19, 6.7**

Amos, H.D., Lang, A.G.P. *These Were The Greeks*. Pennsylvania: Dufour Editions, 1996. **Figs: 1.4**

Army Times, The. *A History of the U.S. Signal Corps*. New York: G.P. Putnam's Sons, 1961. **Figs: 9.5**

AT&T. *Principles of Electricity*. USA: AT&T, 1961. **Figs: 2.24**

Barry, Roger G. and Chorley, Richard J. *Atmosphere, Weather and Climate*. New York: Routledge, 1998. **Figs: 7.1**

Beauchamp, Ken. *History of Telegraphy*. London: Instituition of electrical Engineers, 2001. **Figs: 8.3, 8.5**

Berger, Melvin. *The National Weather Service*. New York: The John Day Company, 1971. **Figs: 9.7**

Born, Max. *Atomic Physics*. New York: Dover Publications, Inc., 1969. **Figs: 2.6, 4.5**

Branson, Lane K. *Introduction to Electronics*. New Jersey: Prentice-Hall, 1967. **Figs: 2.17, 2.19, 2.23**

Bolt, Bruce A. *Earthquakes and Geological Discovery*. New York: Scientific American Library, 1993. **Figs: 7.5**

Brown, Lloyd A. *The Story of Maps*. New York: Dover Publications, Inc., 1977. **Figs: 10.13**

Cartledge, Paul. *Democritus*. New York: Routledge, 1999. **Figs: 1.5**

Coe, Lewis. *The Telegraph/A History*. NC: McFarland & Co., 1993. **Figs: 8.4, 8.6**

Crowther, Arnold James. *Ions, Electrons, and Ionizing Radiation*. New York: Longmans, Green and Co., 1924. **Figs: 2.12**

Danielson, Eric W., Levin, James, and Abrams, Elliot. *Meteorology*. New York: McGraw-Hill, 1998. **Figs: 5.2, 6.9, 10.16, 10.17**

Davies, Paul. *The New Physics*. Cambridge, Ma.: Cambridge University Press, 1989. **Figs: 2.20, 2.22**

Doviak, Richard J. and Zrnic, Dusan S. *Doppler Radar and Weather Observations*. San Diego: Academic Press, 1993. **Figs: 10.12**

Flemming, Roger James. *Meteorology in America, 1800–1870*. Baltimore: The Johns-Hopkins University Press, 1990. **Figs: 9.2, 9.3, 9.4, 9.6**

Gordon, Adrian, Grace, Warwick, Schwerdtfeger, Peter, and Byron-Scott, Roland. *Dynamic Meteorology*. London: Hodder Headline Group and New York: John E. Wiley & Sons, 1998. **Figs: 5.11**

Gray, Harry B. *Chemical Bonds: An Introduction to Atomic and Molecular Structure*. Sausalito: University Science Books, 1994. **Figs: 4.1, 4.7, 4.8, 5.6, 5.7**

Hargraves, J.K. *The Solar-Terrestrial Environment*. New York: The University of Cambridge, 1995. **Figs: 7.13, 7.14**

Herzberg, Gerhard. *Atomic Spectra and Atomic Structure*. New York: Dover Publications, 1944. **Figs: 2.8, 2.9, 2.11, 2.14, 2.21**

Huffman, Robert E. *Atmospheric Ultraviolet Remote Sensing*. Boston: Academic Press, Inc. San Diego, 1992. **Figs: 10.7**

Mendes-Hussey Graphics. **Figs: 2.2, 2.3, 2.15, 3.4, 3.5, 4.3, 5.12, 6.1, 6.4, 6.5, 8.1, 10.8, 10.9, 10.18, 10.19, 10.20**

Jacob, Daniel J. *Introduction to Atmospheric Chemistry*. New Jersey: Princeton University Press, 1999. **Figs: 5.5**

Joos, Georg and Freeman, Ira M. *Theoretical Physics*. New York: Dover Publications, Inc., 1986. **Figs: 2.10**

Kearey, Philip and Brooks, Michael. *Introduction to Geophysical Exploration*. London: Blackwell Science Ltd., 1991. **Figs: 7.6**

Kidder, Stanley Q. Vondar and Haar, Thomas H. Vonder. *Satellite Meteorology an Introduction*. New York: Academic Press, 1995. **Figs: 10.1, 10.5, 10.6**

Kihn, Karl F. *Basic Physics*. New York: John E. Wiley & Sons, 1996. **Figs: 2.4, 2.5**

Leveque, Pierre. *The Birth of Greece*. New York: Harry N. Abrams, Inc., 1994. **Figs: 1.2**

Lerner, Rita G. and Trigg, George I. *Encyclopedia of Physics*. New York: VCH Publishers, 1991. **Figs: 2.16, 4.4, 7.9, 7.10**

Lutgens, Frederick K. and Tarbuck, Edward J. *The Atmosphere*. New Jersey: Prentice Hall, 2001. **Figs: 3.1, 5.9, 5.10, 6.3, 6.8**

Masterson, William L. *Chemical Principles*. Philadelphia: W.B. Saunders Company, 1996. **Figs: 2.1, 2.7, 4.6**

McCormac, B.M. *Magnetospheric Particles and Fields*. Dordrecht, Holland: D. Reidel Publishing Co., 1976. **Figs: 7.11, 7.12**

McDonnel, John J. *The Concept of the Atom from Democritus to John Dalton*. United Kingdom: The Edwin Mellen Press, 1991. **Figs: 4.2**

Monmomier, Mark. *Air Apparent*. Chicago: University of Chicago Press, 1999. **Figs: 10.14, 10.15**

National Portrait Gallery, Smithsonian Institute. **Fig: 8.2**

Piexoto, Jose P., Oort, Abraham H. *The Physics of Climate*. New York: The Institute of Physics, 1992. **Figs: 5.14, 5.15, 6.10, 10.4**

Pomeroy, Sarah B, Burstein, Stanley M., Donlan, Walter and Roberts, Jennifer Tolbert. *Ancient Greece*. New York: Oxford University Press, 1999. **Figs: 1.1, 1.3**

Ramsey, William L. Phillips, Clifford R. Watenpaugh, Frank M. *Modern Earth Science*. New York: Holt, Rinehart and Winston, 1965. **Figs: 2.13**

Reiter, Elmar R. *Jet Streams*. New York: Doubleday and Company, Inc., 1967. **Figs: 10.21**

Salby, Murry L. *Atmospheric Physics*. San Diego: Academic Press, 1996. **Figs: 5.1, 5.3, 5.4, 5.13, 5.16, 5.17, 6.2**

Sauvageot, Henri, *Radar Meteorology*. Boston: Artech House, Inc., 1992. **Figs: 10.10**

Seinfield, John H., Pandis, Spyros N. *Atmospheric Chemistry and Physics*. New York: John E. Wiley & Sons, 1998. **Figs: 6.6, 7.2, 7.3, 10.2, 10.3, 10.11**

Smith, Phyllis. *Weather Pioneers*. USA: Swallow Press/Ohio University Press, 1993. **Figs: 9.1**

Tannoudji, Claude Cohen, Roc, Jacques Dupont, and Gryberg, Gilgert. *Atom-Photon Interactions*. New York: John E. Wiley & Sons, 1992. **Figs: 7.4**

Turcotte, Donald L., Schubert, Gerald. *Geodynamics*. New York: John E Wiley & Sons, 1982. **Figs: 7.7, 7.8**

Watts, Alan. *The Weather Handbook*. Dobbs Ferry, NY: Sheridan House, Inc. 1999. **Figs: 6.11**

The Charting of an Atmospheric Environment

1
The Establishment of Awareness

Atmosphere

We on this planet have unique knowledge compared to other life-forms. That is, we are aware of an existence. Taken further, the continuation of that existence can be defined as life, and thus a mechanism must be in place to enable that continuation process. In our being, that mechanism is called breathing. All other functions that can be associated with that mechanism I consider support. They are, but not limited to, eating, sight, and hearing, all of which comprise the sensory depth that stimulates an awareness to breathe.

The dictionary defines breathing as an act of respiration or inhaling and exhaling a small quantity of air. That is to say we bring into our selves and retain in an organ designed for that purpose components of a small quantity of atoms in gaseous form or a fluid atomic structure that is compressible. The major atomic structures are nitrogen (78 percent) and oxygen (21 percent), respectively. The balances of the gas structures are formed out of small percentages of argon (.93), neon (.0018), helium (.0005), hydrogen (.00006), and xenon (.000009) (Ahrens, 1994:3).

The oxygen atomic structure previously mentioned which was part of the total inhale quantity, is transferred into the blood for dispersion in order to maintain the taking of another breath (intake of gases, i.e., living) and during the exhale portion a quantity consisting of the carbon dioxide structure is exhausted back into the atmosphere. Because of the importance awareness has to the existence of life and the importance of breathing, which provides the mechanism behind the initiative and exhaustive repetitions, a closer look at where those quantities of atomic structures are obtained and how we observe their presence around us should be considered.

Any search usually starts with what we can readily obtain about an object from exploring awareness of the subject and its relationship to our senses. We can feel something around us although it is not normally seen by the human eye; we know something is there because formations appear to float in it, and sometimes colors of red and blue can be seen when we look outward and upward, depending on where the sun is. As we stand on the ground, the perceptiveness of this "something" (height above ground) cannot be visibly determined. At night, particularly on a clear night, does this statement have meaning?

What is this "thing" that without its presence we will perish? That is to say, a quantity of this invisible something is required to be present when we breathe and must be inhaled for our existence to be maintained. Also, what is the uniqueness about its clarity and various colors?

The particular medium associated with breathing is termed *air,* and through the quantity of its molecular structures a density can be established. Consequently, the boundaries imposed by physical law on this density in proximity to any planetary surface will then establish a mass, and in the Earth's case this particular mass is called the atmosphere.

Early humanity was not aware of the atmosphere or the various components that defined its existence for them because they did not have the benefit of defined states of reference. To that end, the desire to expose and define that existence provided the impetus to understand it, and coupled with retention capacities early man could remember at least some specific experiences that would later be used to establish a relationship and its relative importance to us.

The application of man's interest was fundamental in our progression toward the ability to comprehend an existence, through the retention of an awareness delineating a specific period of time. With this as a start, and at some point in time, the effort was made to identify common periods of recurrent experiences that would delimit an existence of which he could be aware of, besides himself. The question is now "how far back in time will historical records go?" establishing that moment in man's past that indicated he was aware of an experience that defined an invisible existence without which he would perish.

Very early records indicate that man was aware of an existence although not capable of any advanced molecular analysis as we now know it. One of the earliest recorded writings identifying that an existence was in effect came from the Book of Genesis, chapter 1, verse 26, in which is stated ". . . and over the fowl of the air." Previously, the word *firmament* was used to describe an existence in atom form and the words *open firma-*

ment (Genesis 1:20, *Holy Bible*, 1978) to describe the location of an existence that would support the fowl to "fly above the earth" in proximity to the terrestrial surface. Proceeding further, the definition becomes more specific in detail. The "open" firmament is presented to describe where fowl may be found, and eventually the fowl are refined to "fowl of the *air*" (*Holy Bible,* 1978:2).

The next logical step is to consider the path that this word might have taken in centers of reasoning in the continuation of its eventual analysis and where those centers became located beyond the book of Moses. Although Moses was in existence until approximately 1405 B.C. (www.bibleview.org/en/modules), the eventual migration of those peoples was to the west around the Mediterranean Sea, along the shores of what we know today as Syria and Turkey. The natural investigation concerning this word *air* once more becomes important in the antiquity of Greek philosophers. On or about 625 B.C., true enough, there were other linguistic influences that might have defined that medium of existence; however, the terms of our language such as *air* and *atmosphere,* and the particle analysis that accompanies such definitions were structured by the Greek culture, and later their research and conclusions became linguistically adapted to our language through the religious efforts of Muslims after the fall of Rome and later Christian monks' translation of Aristotle's additional writings into Latin text during the twelfth and thirteenth centuries (*The History of Weather,* 44).

The path to our recent understanding from 2,000 years ago covers quite a period. Although many apply the thinking that it was the Roman Empire that provided the mechanism of transport to that philosophical circle in a preconceived definition concerning "air" and the science of it, based upon this area of research, the first rationally focused thought into this physical world came from the Greek civilization. It was this cultural sphere that provided the mechanism that drove the impetus within the philosophy and pseudo-scientific community relative to that understanding. The Roman Empire, although having their share of deep thinkers, concentrated their discussions on other planes such as agriculture, politics (administration), law, art, oration, and medicine (Nardo, 1998:89). The Greek philosophies seemed too complex in their thinking and undisciplined in a speculative understanding for the ordered and administrative Roman mind. The Roman educational system adapted an encyclopedia for their understanding of the sciences at that time, and within that body of writing, although the sciences were identified, the explanation of them was without any theoretical understanding. Therefore, the end result for at least Rome's efforts became more inaccurate with each revision. Since we possess accurate data on the definition of *air,* our search for the correct path of understanding toward the identity of the term must have proceeded from a different cultural route during the prevailing philosophical era.

The leading sciences of that day centered around the concepts of deity worship. In that era, prior to 650 B.C., most all of the temporal existences were found to be under the control of a deity. That is, Zeus, Athena, Hades, Demeter, and the like. This particular thinking was primarily located in and around the Aegean Sea, where the Greek nation began as a combination of several cultures.

To identify Greek culture requires an understanding of its origin. Of course this investigation could travel back in time to some ancient era; however, to do so would lead into a more historical endeavor than would normally be expected in a scientific investigation. Yet science itself does have a family tree of sorts. That is, science itself is defined by the language that its originating culture provided in order to define experiences of previously unidentified phenomena.

In Search of Science

In the case of investigating the Greek origins, the appropriate starting point in time appears to be no earlier than 7000 B.C. or around the Aegean Neolithic Period, considered to be characterized by historical evidence related to the Aegean Bronze Age. This culture that we can identify as the first true Greeks, which now represent the current Greek-speaking culture, first appeared about 2000 B.C. They occupied a plains region that can be located from the Carpathian Mountains to the Ural Mountains, east of the Caspian Sea (Abrams, 15). Historical literature within the writings of Homer identified them as Argives; however, the name we associate with them is Achaeans.

Since the Greek people we currently associate that culture with live in and about the Peloponese and its associate, mainland Greece (fig. 1.1), some form of migratory relationship to those peoples occupying the plains area in Europe about 2000 B.C. should be established. A geoterrestrial investigation could then identify and point to that occupation as being the originators of this present-day civilization. With such a migratory focus,

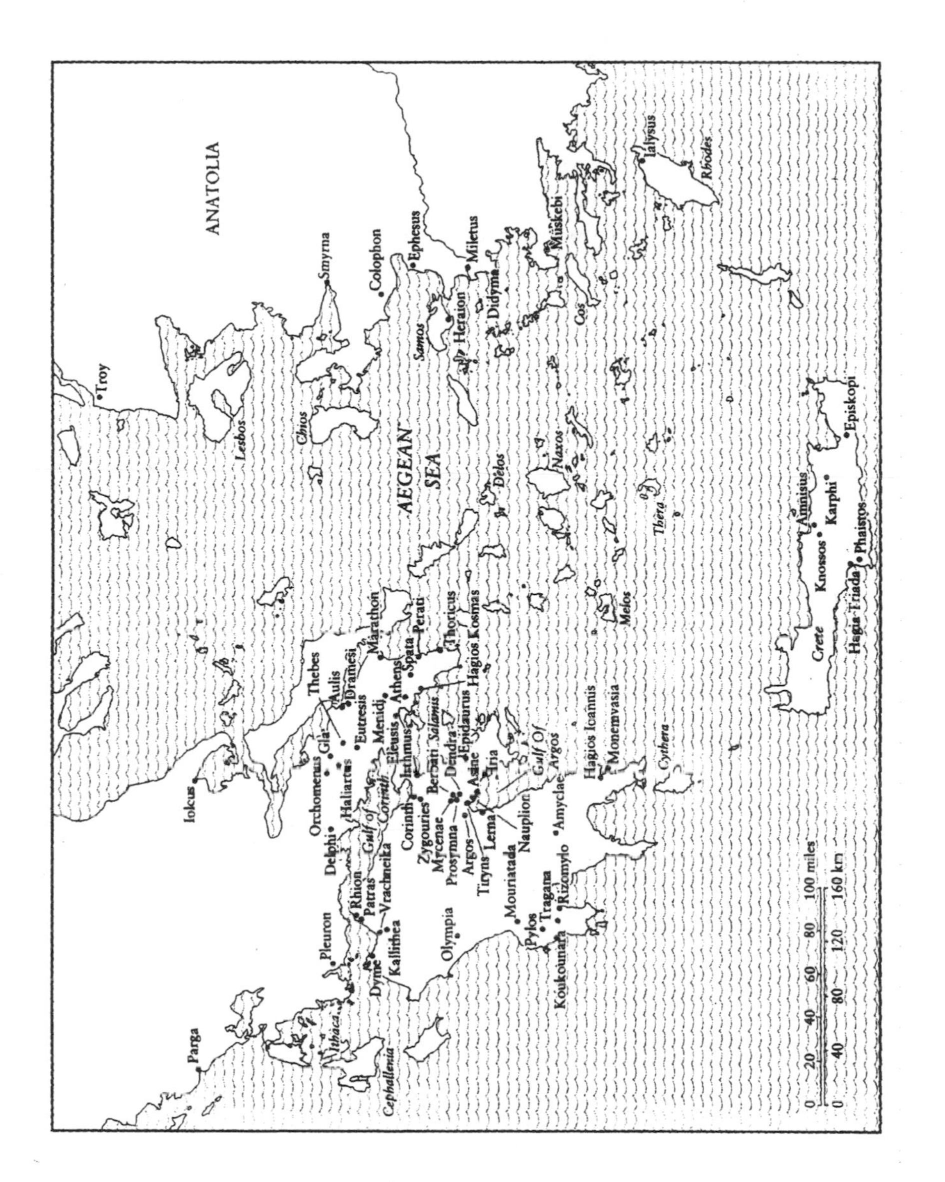

Fig. 1.1

Mycenaen Sites (Thirteenth Century BC.)

(*Acient Greece:* 1999-26)

the culture and direction that established the truest Greek thinkers regarding the origin of certain science philosophies encompassing our investigation of the term *air* may be uncovered.

We can always ask ourselves "what" are the Greeks? An answer in current specifics would then be inclined to define them from a geological position, rather than in a more anthropological one. Typically the answer we would expect would be: They are a representative culture living on extensive landmasses beginning at a point progressing north, consisting of the island of Crete as the southernmost border on the Mediterranean Sea and having its western shore located on the Ionian Sea. The northern range of Greece lies at and borders the northern tip of the Aegean Sea. The easternmost side of this country is located on the western Aegean Sea coastline, consisting of an island chain from Samothrace to the island of Rhodes, including the islands pertaining to the Cyclades. By all accounts, the Grecian landmasses can be said to enclose the Aegean Sea.

If, on the other hand, we ask, "Who are the Greeks?" then some pause is to be expected. A people of today would not necessarily be pure in origin when traced back several millennia B.C., but a composite of anthroprogenics from some migratory process eventually formed a specific cultural identity. To start, the name Greek as listed in the dictionary pertains to a member of one of four tribes called: Achaean, Aeolian, Dorian, and Ionian and later the identifier of that composite. These people at that time B.C. called themselves Hellenes; and their country, Hellas (Adams and Long, 1996:4). The end, however, is not here. These people came after predecessors, that is, the Minions and Myceaeans. The investigation into pre-Greek identity is therefore necessary to trace the first application of scientific philosophies to interpret an origin of certain phenomena. In any exposition of philosophical origins will be found the specific people within that cultural composite initiating a conception, which eventually materialized as a cultural foundation stone.

Ancient Civilizations

The Minion civilization flourished and reached a peak in its culture between 1650 and 1450 B.C. during an era called the Bronze Age, before their culture was replaced by the Myceaneans (fig. 1.2). Evidence suggests that the Minions settled in Asia Minor, Rhodes, Melos, Thera, the

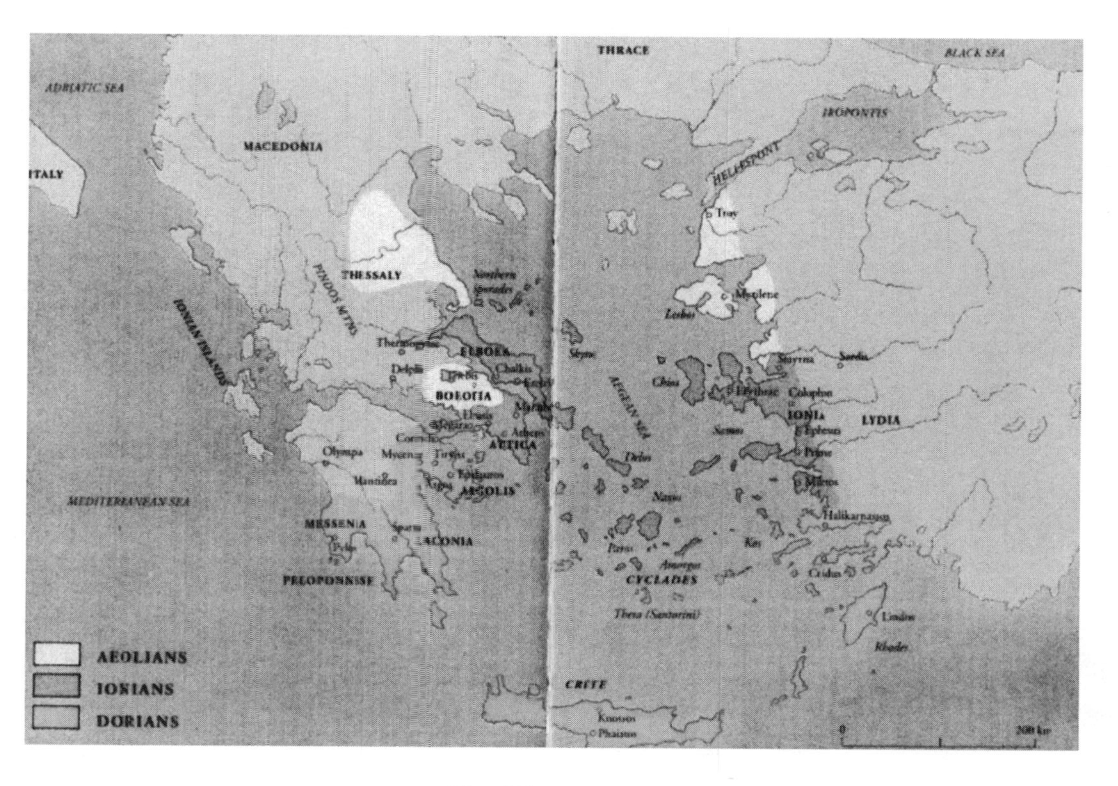

Fig. 1.2
Early Pre-Greek Culture Factions
(Abrams: *The Birth of Greece*)

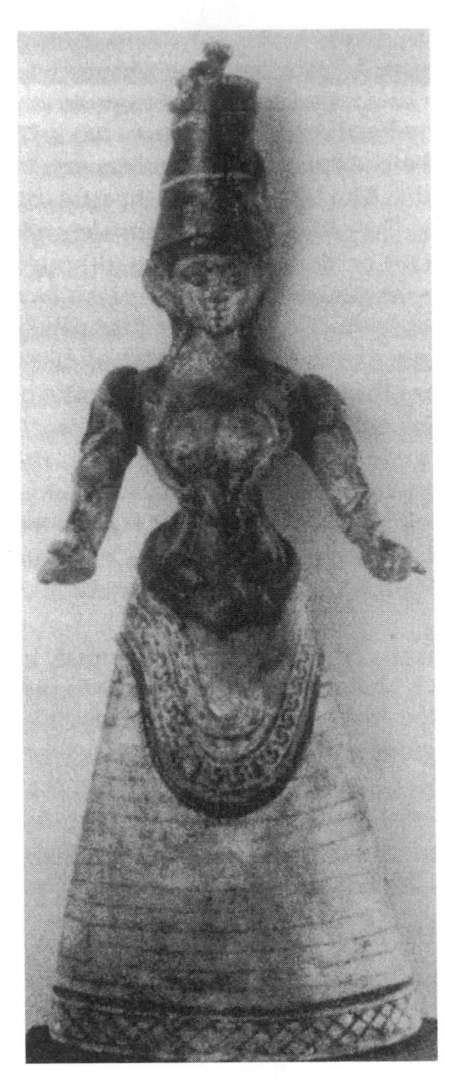

Fig. 1.3
Knossos goddess statuette
(*Ancient Greece*:1999-33)

Peloponnese landmass, Thebes, Athens, and Crete, with the capital identified as the first royal palace, constructed around 1700 B.C. (1999:13) at Knossos. Evidence of their existence can be found in archaeological sites located in and around the areas mentioned. Artifacts that identify this civilization have been found to include pottery, weapons, tools, and jewelry. The Minions possessed a form of writing similar to hieroglyphics, which around 1800 B.C. was adapted to represent signs ("Linear A") that formed syllables. The next step in that communication evolvement was the application of the syllables to the formation of words (Pomeroy, Burstein, Donlan, and Roberts, 1999:13). Their art was depicted as joyful and without a warlike or aggressive nature in representations depicting many forms of social life. As much as this culture flourished in the arts, however, scientific records attempting to identify unexplained physical occurrences seem lacking.

Around the approximate time line of 1500 B.C., a people known as the Achaeans replaced the Mycaeans, whose civilization eventually phased out about 1200 B.C. Archaeology can confirm the type of culture represented through Mycaean artifacts and the Linear B tablets. Their society seems well covered in the inclusion of carpentry, masons, goldsmiths, leather working, and so on. A large textile factory was in operation, which can be supported

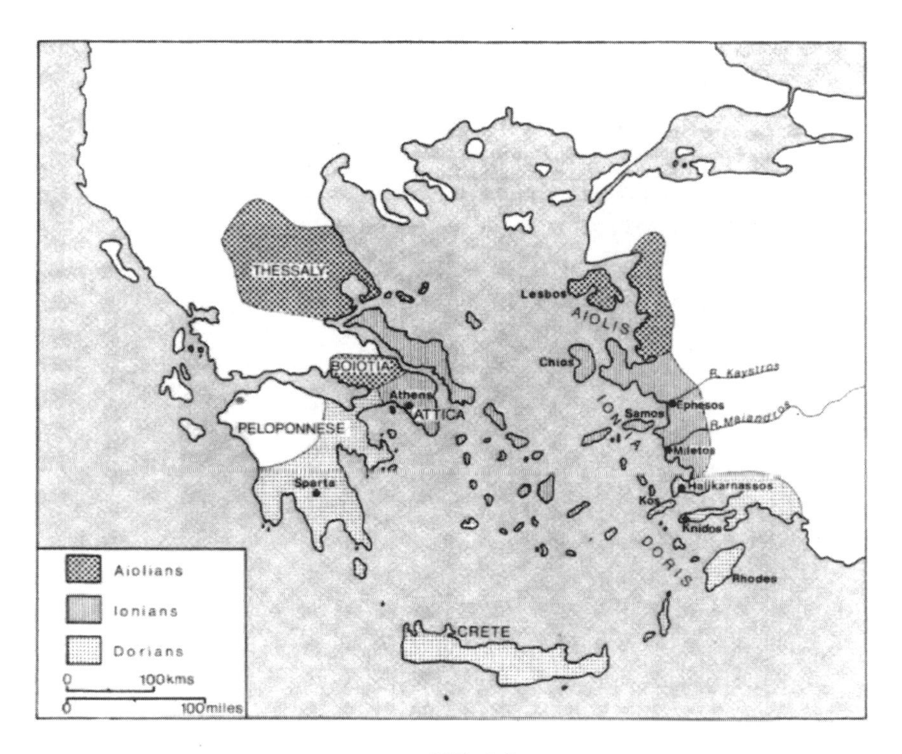

FIG. 1.4

Greek Migrations onto Aisa Minor

(These Were The Greeks: 1996-27)

by additional descriptive tablets at the Knossos Palace (1999:31). Again, however, their pantheist religion replaced science for explanations regarding daily life situations that needed a fix. Mycaean pantheist gods included but were not limited to Zeus, Hera, Poseidon, and Athena (fig. 1.3). From this evidence, another dead end may be concluded to explain any rational and cohesive underlying scientific explanation for physical phenomena.

Archive evidence at the palace of Knossos identifies an Achaean (Greek) prince in the process of ruling there about 1500 B.C. (34), but that the Acheans were responsible for the Mycaean culture's demise after this period is somewhat unclear. The island of Crete was not the only aggressive takeover by the Achaeans; it is also likely that the Achaean presence was accomplished in similar manner along the eastern coast of the Aegean Sea. The Achaeans began their migration about 1050 B.C. off the Peloponnese from the prospects of eventual encroachment by the Dorian faction coming by way of a northern advance into the Peloponnese. It is this eventual Dorianic movement that brought the Achaeans migrating from Athens, and used as an inclusive term to represent the eventual Grecian kingdom, onto the Asia Minor coast orientated in a north–south manner. These refugee settlements eventually stretched from Hellespoint to Rhodes, and because of the tributary arrangement from the rivers of Kaystros and Maiandros (fig. 1.4) connecting the interior to the sea, the region particular to the Ionian civilization became extremely affluent and civilized (1996:26).

The original settlements of the now-displaced Peloponnesian culture migrated onto the mainland area from the western island chain consisting of: Lefkada, Kefalonia, Zakynthos, and Ithal, located near the Ionian Sea's southeastern shoreline. These islands represented a grouping of the chain not really influenced by the Dorian occupation of the Peloponnesus, mostly due to their remoteness from the mainland.

The original occupation of this Ionian island chain began within the Middle Paleothic Era, dating 50,000 to 32,000 years ago (1999:6). It is therefore quite possible to conclude that since Mycaean evidence is rare on the island groupings, which may indicate a lack of established presence (13), that the original inhabitants on the Peloponnese, through occupation by the Minions, then Mycaeans, then Acheans, and eventually the Dorians, may have been Ionians already there in a settlement established by migration off the islands many years prior.

The settlements along the Asia Minor coastline eventually became arranged in a north–south line, the north by the Aeolians, the center by the Ionians (which may have been the eventual Achean migration from the Peloponnesus). The southern Asia Minor coastline starting south of Miletus to the island of Rhodes, westward to Crete, and upward to the Peloponnese was occupied by the Dorians (fig. 1.4).

Interesting enough was the lack of rational thought regarding environmental explanations through these periods. There was still a pantheistic viewpoint evidenced by their continuing belief in many gods in control of everything. The Achaeans also succumbed to the new Mycaean deities when they took over the island of Crete and continued to use idols that promoted solutions through their own individual religious expectation. These forms of nonscientific explanation were still in existence around 800 B.C.

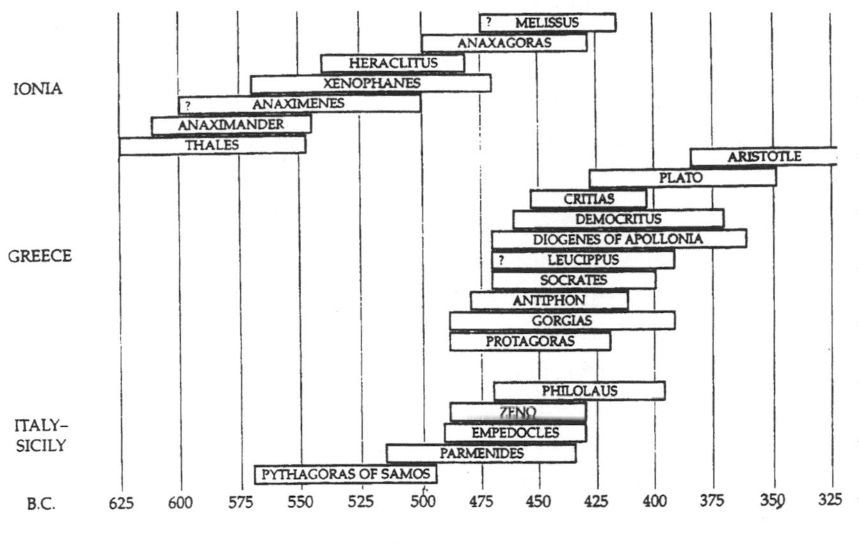

FIG. 1.5

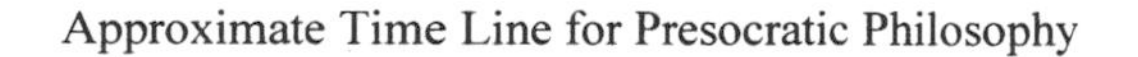

Approximate Time Line for Presocratic Philosophy

(*Democritus*: Cartledge:1999)
(no page identified)

Since the Ionian settlements on the Asian coastline were considered centers of great exploration and trade, it is perhaps possible that the answer to our question identifying an origin of philosophies applied to physical explanations may exist in this line of exploration, as such, influential factors in the promotion of expansive thinking often manifest themselves as a byproduct of cultural assimilation. Further exposition of philosophies around this period with a focus on the Ionian presence may lead us in the appropriate direction.

The direction would logically be to identify a source of rational thought associated with a structured logic. Since the area of our expositional structure of rationale is science and within the science are contained undeniable laws regarding physical worlds, then perhaps early philosophers in this region in their approach to determine early explainable laws may have considered discussions into areas where an application of laws had not been brought to bear.

Pre-Socratic evidence of the eastern Aegean (fig. 1.5) contains several individuals that could be focused upon as being Ionian and from that particular coastal region around 600 B.C.

A Philosophical Foundation

I believe that in the search for these individuals perhaps a study into what may have caused such an outburst within their society at large should be undertaken. This direction will assist greatly in the understanding of why their philosophical thinking came to be exposed.

First in our search we should identify those areas of discussion within the Ionian society that formed the basis of any thinking that was considered contrary to established norms, medicine with an Egyptian influence, arts and the Near East influence, astronomy (Egypt), and the Babylonia calendar (Loyd, 1970).

These scientific areas that were once determined to be under some deity, combined with the advances in government and political debate, established an inquisitive posture. Open discussions became more focused on seeing the reality of a subject determined by the advancement of provable concepts rather than an explanation that some deity did it or only a deity knows and therefore the matter was closed.

Observations beyond the sociopolitical were presented within the debate structure as would either a law or other sociopolitical consideration. This forum environment tended to cement together various postulations, eventually drawing a consensus from the previous unenlightened. Therefore, various considerations presented for debate from such a diverse assembly contributed to an eventual outcome that could be identified as a form of rational filtration. The Ionian city of Miletus was foremost in this form of thinking structure, and, individuals of certain pre-Socratic philosophies eventually became directed toward an awareness of the natural environment rather than only implications regarding a law or social judgment. These exchanges relative to temporal observation, when presented in a forumlike setting for that era in an attempt to explain an observation that was not deity-originated, were controversial, to say the least.

Through the advances in science comprising medicine, mathematics, and astronomy more and more of the deified world was being proved obsolete. From this impetus, a new order of thought with regard to nonempirical exposition came into existence. One of the earliest Ionian pre-Socratics most recognized in establishing this form of expositional philosophy was Thales of Miletus (approximately 625–547 B.C.).

Although terms that related to actual verbiage such as *matter* and *substance* were not used in the sixth century B.C., the discussions of origins were (Lloyd, 1970:18). On the principles of matter, Thales believed that water was the most probable beginning of such things (1983:89). This form of "beginnings" concept was not generally considered by the philosophers of that time with the exception of Thales, the founder of the Ionian School. They presented a consideration whereby all things come from only one source and change in qualities from inception to end. Therefore, an absolute beginning of multiple substances could not exist.

As far as the Earth or a substance was concerned, Thales presented a consideration of his own. He felt that the Earth was supported by or floated on water and that the water was indeed the basic "constitute of all things" (Barnes, 1983:91). Accordingly, some reviewers of the understanding of that era explain that Thales's using water as his basis of all things was perhaps the rational result of the astronomical importance of that day and the quasi-deified thinking that era promoted. The fact that Thales was still considered to be a recognized astronomer, I believe, played no small component in the administration of his precedent thinking. In the prediction of not only a solar eclipse but also solstices there was a demonstration that an order of expectations could exist and not a random sequence or activity that would have been attributed to be the product of a deity (1999:122).

Thales was certainly of a logical mind in his attempts to discern the natural state of observable natural phenomena. His ability to factor mathematics in the deducement of the height of the pyramids, at least according to their shadow, was documented by Hieronymus of Rhodes (1983:85). Three other theorems are attributed to Thales by Proclus regarding dimensional geometric calculations. One is that the circle as bisected by its diameter, the second regards the base angles of an isosceles triangle being equal, and third is that the vertically opposed angles are equal. Again his contemporaries would have respected these mathematical explanations without empirically based foundations. In other realms of thought beyond the mathematical, Thales did initiate debate around the subject of "substance" and "matter," most profoundly that of water and later through his contemporaries that of air. He, for the most part, provided a stimulus for further philosophical discussion regarding the basis of all things. Although there is no real evidence that Thales wrote any of his theories down, as substantiated within the Alexandrian library, early Greek writings by others depicting Thales's recognition exist.

With this as a start toward expositional rationalism, we still must not forget that in this era of pre-Socratic pattern of thought, the Milesians postulated without empirical support. (They debated their thinking without actual experimentation, thus promoting among themselves an inductive theory that could explain observations only if a consensus was achieved and if no modeling data was available) (1982:49). The infusion of this thinking process into pre-Socratic philosophies continued the rational exposition into a more scientific realm by additional expansions relative to the beginning of all things. The next search of continuing philosophies within the Greek postulate structure will define a further evolution toward the establishment of a material foundation (fig. 1.5) as presented for debate, by another consecutive Ionian: Anaximenes of Miletus.

Anaximenes of Miletus (600–528 B.C.), a pupil of Anaximander and the last of the Milesian School, replaced the water idea, or the medium (Apeiron) as termed by Anaximander, in which the Earth was held in place or floated by equally strong forces that were produced by all heavenly bodies. It is this term that Anaximenes replaced with another: *air,* which he felt described something without form and yet boundless in dimension. The Greek word for the air we understand to be was spelled *aer.* The people of that time could feel something around them and although they could not actually see this substance, they knew that whatever it was must be controlled by some deity. The Greek god of the air was Aloes, son of Poseidon and Melanippe (Burr, 9) and considered the highest lord of wind and air. Anaximenes was familiar with the Greek gods regarding the various influences subject to them and chose instead to apply a rational explanation to a substance that was considered under deified control. It is also intriguing that the term *air* was used within the Book of Genesis, which depicted the region of air as the open firmament of heaven, as created on the fourth day. In the Bible, the term *air* was used regarding where the fowl would fly. These transferences would have originated from the Hebrew Old Testament and passed to the Greek civilization, possibly from the Egyptians through trade and/or through the diaspora (Haywood, 1997:27). Although the Scriptures themselves would not be written for another 100 years or so, it is evident that Anaximander was aware of some form of temporal creation regarding *aer* (air) and Hebrew record. It is perhaps the contemplation of the creation of this invisible substance where the fowls fly and the dynamic properties that could be felt that prompted the linkage between them.

The rational idea that clouds and water seemed to originate and become a substance from this invisible something (Popper, 1998:36) could not be physically defined by dimension. This "air" was a reality as far as its temporal properties went and could be felt through a range of dynamics or wind as well as the experiences of heat, cold, wetness, and dryness. In this process of thinking, however, theoretical it may be, there was always the fact that whatever the origin was, it changed into something else. Clouds came from nowhere, heat became cold, dryness turned into wetness, and so on, which was at least an observation of the natural process. Interestingly enough, although these processes were accepted generally within the pre-Socratic philosophies as observable phenomena, it was Anaximenes who attempted to unify an explanation within the bounds of a single material. Also, when single material morphed into other forms opposite to and in contrast to each other, that alteration, he reasoned, could be explained by its density (Barnes, 1983:146). It was his argument that as the air became denser from its invisible (rare) form, the reaction was to become something else that became visible. In his terms, the appearance became observable by a change that depended on how much there was of this quantity in a certain place.

Even with this argument, an underlying thesis remained. Because a pre-Socratic fundamental philosophy was to identify the beginning of all things, it was here that this presentation became difficult. If one basic identity changed into another and there was no retention of the original within the new form, then the origin material could not be the elementary foundation and must still be hidden. Therefore, there must be a difference between a temporal reality and the appearance of it.

Although the corporeality of air was observable in many ways, Anaximenes was at least the first to initiate the concept of explanations of materially atmospheric terms and, more important, in the direction of its invisible existence, at least to our senses. From this standpoint of rationality, the eastern side of the Aegean Sea, specifically Miletus, came to a halt and fell into a period of inactivity. Further association with the term *air* was moved into the background and replaced with a more accurate and real identity from the Greek philosophers of Leucippus and Democritus. These individuals presented the next refinement to complete our current quest for atmospheric rudimentation.

The Invisible Defined

Leucippus (485 to 390 B.C., estimated, fig. 1.5) was from Miletus as well. An association with Parmenides, at least in the philosophical arena (1957:402), was the extent of Leucippus's agreement with his contemporaries. He was considered to be an Ionian and a pupil of the philosopher Zeno. By that expression, Leucippus identified realities around him in a more physical way, unlike the definitions postulated by Xenophanes and Parmenides.

Leucippus presented a remarkable consideration for what we now refer to as the truth of the matter. That is, that our reality is composed of the smallest, and by the smallest I mean indivisible, particles (Nardo, 1998:48), also that these particles are varied in their shape, size, and particular formation orchestration and that when these tiny particles come together, a substance is formed. Here were the first steps toward a significant philosophical relationship with reality toward the primary basis for current ontological thought, insofar as theoretical thinking and its link with matter are concerned. Consideration was given that although we cannot perceive a thing, the thing is in existence because its size is so small that we cannot sense it and it is therefore imperceptible to us. We can know that this small particle exists because when it combines with others of its kind the particular arrangement causes an enlargement to a point at which we become aware of its reality. The ontology was a philosophy that additionally constructs the smallest particles as having "no void," which made them indivisible, as it was the void within particles that gave them divisibility. The foundation that the void was an existence and could be identified as being independent from the particle is important. Although Leucippus did not identify exactly what the void was, it gained definition by what it was not. Therefore, its existence was predicated upon occupation and not a material substance or particle (Kirk, Raven, and Schofield, 1983:415). The word used to describe the tiny indivisible particle was *atomoi* (Guthrie, 1975:58), which meant "unsplittable," and defined these particles as being indestructible. Motion by way of contact or impact was also presented as a foundation of this ontology. However, the argument arises as to what initiates first motion. Leucippus was not given much credit from his contemporaries in this arena of ideas; consequently as one of the founders of the term *Atomists,* he falls some distance behind the other individual who shares this title, and that man was Democritus.

Democritus was also from the city of Miletus, born around the time between 460 and 457 B.C. and died about 404 B.C. or about the age of 100 years (1999:60). Also a pupil of Anaxagoras and Leucippus, Democritus followed the ontology of Leucippus to the extent that a goodly portion of the Atomists' principles are subscribed to him. He presented his thinking in the sphere of shape and where alteration was a product of arrangement. As to the existence of atoms, his thinking was directed toward the joining or separating of them. When they were of multiple adherence, then matter or substance existed. When they were divided into the singular, it did not. Interestingly, the existence of an object was in the appearance of it. If the object was separated in atomic form to its basic unity or atom (no arrangement), then it had no appearance, therefore no existence. It is important here to note that I believe a difference existed between the ontology of both Leucippus and Democritus. Existence for Leucippus depended upon atomic size. The reason we cannot perceive the atom is its small size. The Democritean Principle takes this ontology differently. His presentation on the existence of the atom was not in atomic size but rather in the adherence of two atoms. This is fundamental monism because only atoms of like material will adhere and that was the result of motion through collision (Barnes, 1983:572).

The philosophical argument that an adherence cannot take place with one but requires two singularities causes the conflict that any existence of one without another must be true or adherence could never exist (Barnes, 1983:406).

There was a skepticism in the way Democritus viewed the reality of atomic structures and in the way we perceive them. He felt that although a substance existed through atomic adhesion, the final decision of the perceived substance was in the eye of the beholder and not a general truth. The wind may have particles and structures that we can be aware of; however, two individuals may perceive them differently. An example would be one individual having the perception of hot wind and another having the perception that the wind is cold. Also, Democritus presented the thought that atoms have weight insofar that the substances they created varied in weight. This was of course attributed to the density ascribed to the variations of spacing between the atoms or how much void existed within the substance structure. The size of the atom was considered to be variable, so an atom could be the size of the universe or imperceptible, which added a finer delineation to the weight aspect.

Other areas of speculation were presented as the reasoning behind color. Democritus's theory of that medium was explained also in terms of atomic characteristics. The color of a substance was due to the color of the particle, and the shape of the atom fabricated the concept that an angular, thin atom produced an acid taste, while a larger atomic size would produce a sweet taste (Loyd, 1970:48). Again these philosophers engaged in conceptual thought without empirical data, and this component of the physics and theoretical approach would certainly hamstring any scientific advancement that would have been predicated upon hard reality.

Here, to this point of our study at least, the origin of the word *atom* has been exposed. Further enhancements to this historical philosophy and ontology to finally transcend into the breathable and identifiable medium still remain.

2

Invisible Potentials

The "Real" Atom

The philosophical word has some definition of a particle that theoretically exists called the atom. A refinement would be necessary beyond the fifth century B.C. to adequately represent current substantiated existence of such a reality and its place within the temporal influence called sphere.

It wasn't until the Italian physicist Gassendi suggested much the same ontology as the pre-Socratics studied. Later still, Sir Isaac Newton (1642–1727) presented much the same support for the atomic theory of Gassendi (1966:30). Years passed until 1800, when the theory of elemental matter having components consisting of small particles was again elevated to a higher conformation of physical thought. In 1808, John Dalton revised that thinking again to include his presentation that not only are all atoms in an elemental substance alike, but that a variation of multiple composites would be formed if the atomic structures of different elements combined in specific proportion to themselves (1966:30).

It was this philosophy that was combined and applied to the ontological theory of the pre-Socratics to which may be ascribed in the present day the term *Atomic Theory.* First, in any exposition regarding the development of theories with respect to the atomist's there is an association that invariably mentions the term *element.* In this case the word means the absolute basic singularity. Therefore, an element is one of a specific and distinctive existence; it may combine with others of different distinction but by itself in that only atoms of a specific existence will constitute the reality called matter in its most indivisible state.

There are three basic postulates with regard to John Dalton's theory of the atom:

1. An element is composed of extremely small particles called atoms.

As an example, hydrogen is an element; it is composed of the atom hydrogen. Another element, oxygen, is composed also of oxygen atoms. In each case, the composites of atoms that create the elements of both hydrogen and oxygen are unique to only themselves.

2. Atoms of different elements will have different chemical properties. In the course of reactions that would alter the composite element, the atoms themselves will not be changed. An oxygen atom will never become a hydrogen atom, and so forth, for elemental atoms.

3. A compound substance is formed from the combination of more than one elemental atom. In a pure compound, all the elemental atoms present will be constant and very definite. As an example of this, water consists of both the element of hydrogen and the element of oxygen and as such is considered to be a compound substance. It is important to state here that a pure compound substance consisting of singular elements can be considered to constitute a molecule (1924:33). As a defined structure there will always be two hydrogen atoms for every one oxygen atom, and this is with regard to the mass of a compound substance (Law of Conservation of Mass).

In an alteration of substances whereby the elements may be separated in a reaction, the total mass representing all atoms present would not change because all the atoms are still present, although not in substance form.

In any pure compound consisting of one or more elements, since the atom count is always constant, the weights that represent the individual elements and their percentages therefore will never change (Law of Constant Composition).

If a compound substance has a multiple combination of elements as in H_20, then the weight ratios would be two times the hydrogen atom and one times the oxygen atom (Law of Multiple Proportions).

The Law of Combining Weights is rather different because of the actual atomic configuration representing the electron orbits with respect to the nucleus. Because of this variation in compound substance makeup, more about valence configurations presented by Frankland in 1852 (1966:38) will be discussed at a later time. Suffice to say that at this point in discussions relative to atomic theory, the previous laws are in practice today.

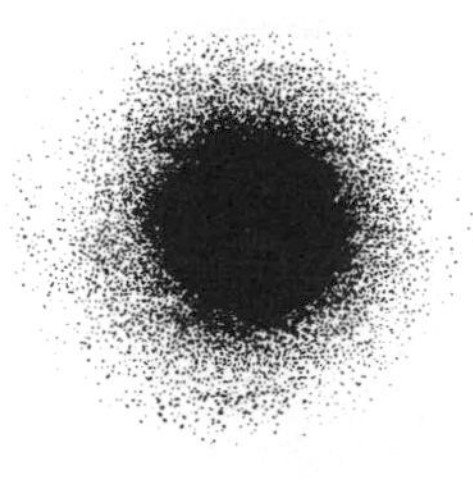

Fig. 2.1
Basic Atom Structure
(Chemical Principles:1966-164)

Atomic Structure

The atom is a fascinating structure that is the basic foundation of all reality. Thus, because reality does exist, there must be some stability in the formations that enables matter to exist for such a long period of time. Investigative means have determined that the atom itself is comprised of other components that themselves have varied positions and paths within the atomic structure (fig. 2.1).

Close inspection reveals a haze about the dark, more solid portion of an atom in figure 2.1. The reason for the haze portion is an activity regarding those other components. In fact, there was considerable speculation regarding the existence of smaller particles than the atom itself. In 1891, Sir William Crooks had observed something through low-pressure gases that would again be identified with the existence of an unseen particle. In 1891, G. Johnstone Stoney proposed that the name for this unit of particular charge be "electron" (Branson, 1966:5). With this discovery and the term *electron,* in 1897 an English physicist named Sir J. J. Thomson in an experiment used a newly invented device that when connected to a high voltage potential, observed what would be later termed *cathode* rays because of the observable bluish light at one end of the device (fig. 2.2).

In the observation of an apparent connecting stream, a deflection of that stream could be initiated by a voltage connection to a polarity and magnetic field, attached to contact plates arranged on opposite sides of the

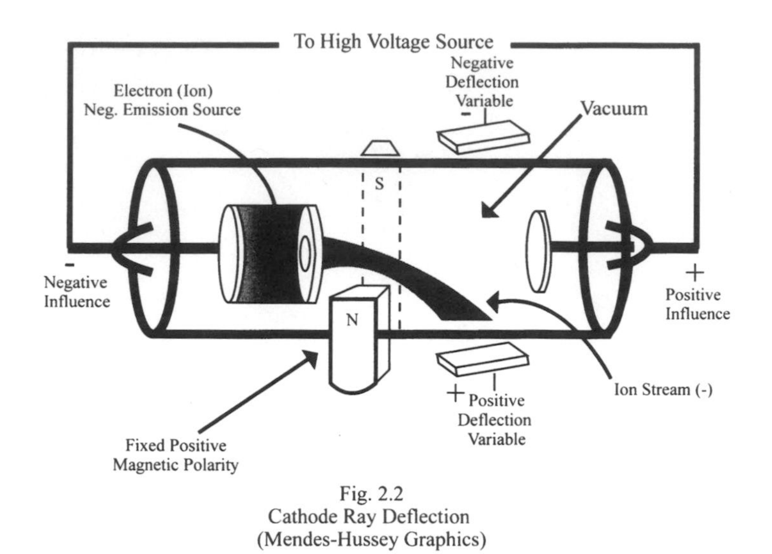

Fig. 2.2
Cathode Ray Deflection
(Mendes-Hussey Graphics)

glass tube (fig. 2.2). It became his conclusion through this observation that the particles in motion must have had some energy force within them to be attracted to the positive plate or, at the very least, have something about the particles that became attractable to other material.

With this observation, his discovery, and associated credit was another page of information regarding the electron particle and to explain a more comprehensive exposition on the atom itself. Up to that point, there had been no information known to exist concerning the electron with respect to specific charge or mass as an individual state. Also, the atom must have a neutral charge by the intermediary of an equally offsetting potential of its additional components. That is to say, if the discovered electron's existence has a potential, then it is reasonable to assume that the "other" components must likewise have an offsetting ability to account for the neutrality of the atomic structure in rendering matter stable. This suggested that a ratio existed within the atom in the number of electrons it might have to satisfy the weight of the mass that it represented in matter.

In the period between 1911 and 1913, three men—Ernest Rutherford along with H. Geiger and E. Marsden—proved that most of the mass of an atom along with a positive potential were indeed concentrated in a very small volume having a radius of less than ten to the minus fourteen meters of the outer radius of the atom (1991:82). This was complemented in 1913

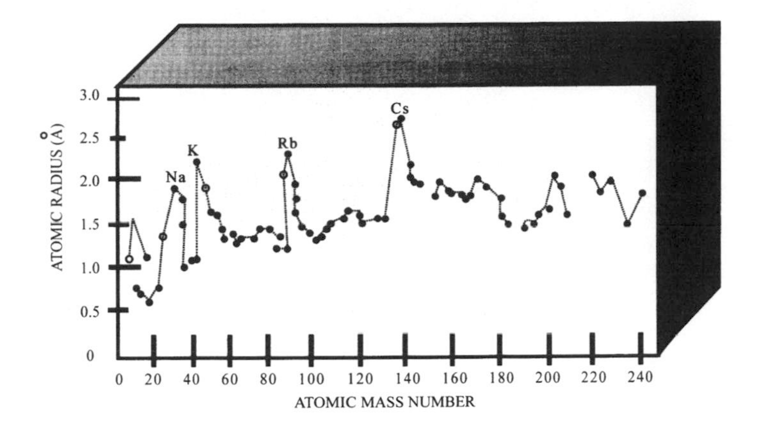

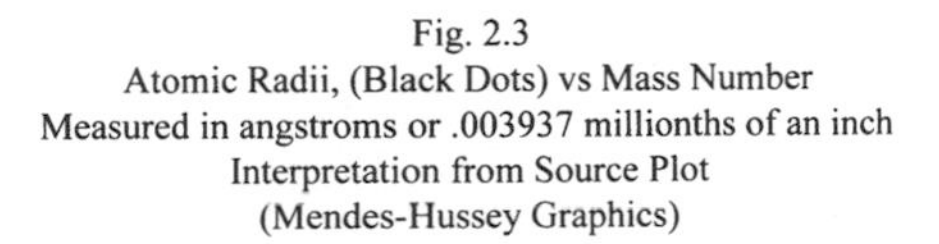
Fig. 2.3
Atomic Radii, (Black Dots) vs Mass Number
Measured in angstroms or .003937 millionths of an inch
Interpretation from Source Plot
(Mendes-Hussey Graphics)

by Niels Bohr, who developed an accurate picture of an atom using hydrogen as the example, with the nucleus having a positive potential (+), associated with 99.95 percent of the atoms' total mass. Circling around this core nucleus in orbit was a single electron. It must be stated here concerning additional orbits or levels added around the inner nucleus of an atom that the size of that atom does not necessarily grow dimensionally (fig. 2.3). One of the first listings of more than the hydrogen element was defined by Antoine-Laurent Lavoiser, who through observations of chemical decompositions and recombinations added about thirty more. Later the proportionality of the atomic weight characteristic (number of protons within the nucleus) was advanced by Dimitri Mendeleev in 1869 in one of chemical history's most important discoveries. He determined that elements have specific properties and that the properties themselves are periodic as a function of their weight (fig. 2.4). In addition, they could be incrementally sequenced into columns and rows. Elements of similar properties would fall into specific categories, showing for the first time that there was an interrelationship among them.

Theoretically at least, thinking was given to just how much an atom would weigh given that additional electrons could be in orbit around its nucleus. After all, if an atom had most of its mass in the nucleus and the nu-

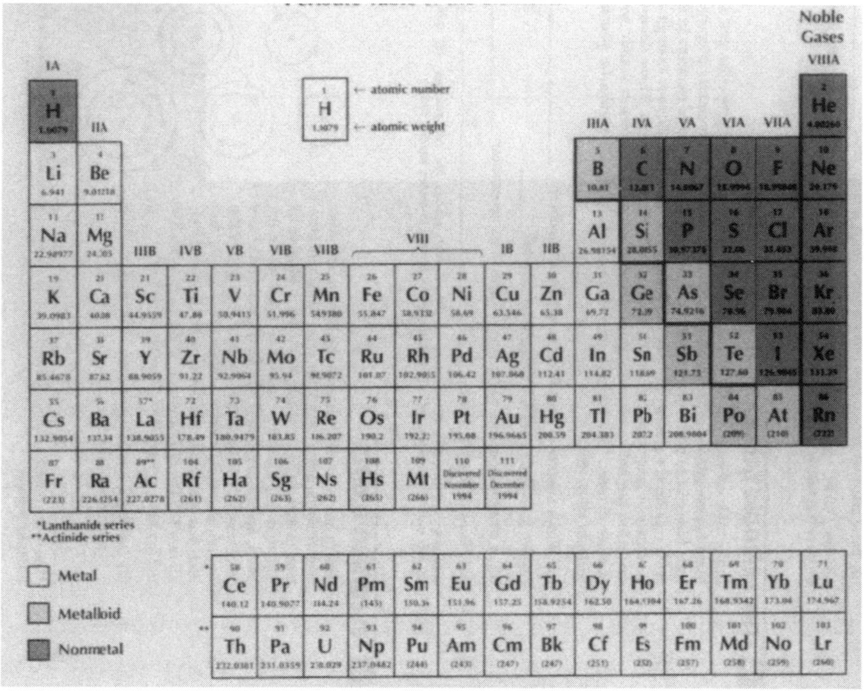

IA	IIA	IIIB	IVB	VB	VIB	VIIB	VIII			IB	IIB	IIIA	IVA	VA	VIA	VIIA	Noble Gases VIIIA
1 H 1.0079																	2 He 4.00260
3 Li 6.941	4 Be 9.01218											5 B 10.81	6 C 12.011	7 N 14.0067	8 O 15.9994	9 F 18.99840	10 Ne 20.179
11 Na 22.98977	12 Mg 24.305											13 Al 26.98154	14 Si 28.0855	15 P 30.97376	16 S 32.06	17 Cl 35.453	18 Ar 39.948
19 K 39.0983	20 Ca 40.08	21 Sc 44.9559	22 Ti 47.88	23 V 50.9415	24 Cr 51.996	25 Mn 54.9380	26 Fe 55.847	27 Co 58.9332	28 Ni 58.69	29 Cu 63.546	30 Zn 65.38	31 Ga 69.72	32 Ge 72.59	33 As 74.9216	34 Se 78.96	35 Br 79.904	36 Kr 83.80
37 Rb 85.4678	38 Sr 87.62	39 Y 88.9059	40 Zr 91.22	41 Nb 92.9064	42 Mo 95.94	43 Tc 98.9072	44 Ru 101.07	45 Rh 102.9055	46 Pd 106.42	47 Ag 107.868	48 Cd 112.41	49 In 114.82	50 Sn 118.69	51 Sb 121.75	52 Te 127.60	53 I 126.9045	54 Xe 131.29
55 Cs 132.9054	56 Ba 137.34	57* La 138.9055	72 Hf 178.49	73 Ta 180.9479	74 W 183.85	75 Re 186.207	76 Os 190.2	77 Ir 192.22	78 Pt 195.08	79 Au 196.9665	80 Hg 200.59	81 Tl 204.383	82 Pb 207.2	83 Bi 208.9804	84 Po (209)	85 At (210)	86 Rn (222)
87 Fr (223)	88 Ra 226.0254	89** Ac 227.0278	104 Rf (261)	105 Ha (262)	106 Sg (263)	107 Ns (262)	108 Hs (265)	109 Mt (266)	110 Discovered November 1994	111 Discovered December 1994							

1 H 1.079 ← atomic number ← atomic weight

*Lanthanide series
**Actinide series

*	58 Ce 140.12	59 Pr 140.9077	60 Nd 144.24	61 Pm (145)	62 Sm 150.36	63 Eu 151.96	64 Gd 157.25	65 Tb 158.9254	66 Dy 162.50	67 Ho 164.9304	68 Er 167.26	69 Tm 168.9342	70 Yb 173.04	71 Lu 174.967
**	90 Th 232.0381	91 Pa 231.0359	92 U 238.029	93 Np 237.0482	94 Pu (244)	95 Am (243)	96 Cm (247)	97 Bk (247)	98 Cf (251)	99 Es (252)	100 Fm (257)	101 Md (258)	102 No (259)	103 Lr (260)

Metal
Metalloid
Nonmetal

Fig. 2.4/2.5
Periodic Table
(*Basic Physics*: 1979-68)

cleus was in proportion to the number of electrons in orbit about it, then its mass must be heavier in compensation and much denser due to the absence of incremental radii. If the atom is to remain neutral, then its nuclear (nucleus potential) charge must be equal to the number of electrons in orbit representing the electron cloud. This nuclear potential was identified by Wien in 1898 and Thomson in 1919, eventually given its name, proton, from the Greek word meaning "first one" (1991:964) by Ernest Rutherford in 1920. Currently, the number of protons within the atomic nucleus is considered to be the "atomic number." In addition, the mass of the total nucleus is represented by not only protons but an additional particle called the neutron, represented in equal numbers with a net nuclear charge of zero. The total sum calculated between the number of protons and neutrons within the nucleus of an atom represents the "atomic mass." As an example of this, refer to figure 2.5.

Principles of Elemental Unification

There is an importance regarding the atom and a relationship to the atmosphere, because there must be more than just an atom consisting of a nucleus with one proton, one neutron, and one electron in orbit around it.

The Greek pre-Socratics felt that atoms were in existence, and we now know that with discoveries up to 1920 they were indeed correct in their thinking that atoms do exist. But how many and what kind? Further, if our reality exists of a variety of elemental compositions, then there must be some way that the atoms (if there are more than just one set of nucleons) combine to form other pure substances. Also, logic would shepherd rational thought toward applications of atomic structures into densities that could be in existence but invisible to us. Further, that such densities could encompass other forms of realities constituting an outer boundary of nucleonic strata might be possible.

This was reinforced by the experiments of H. G. J. Moseley in 1913 (1991:349) observing the K and L X-ray energies that defined each element by a particular characteristic. It was through the experiments of X-ray bombardment that the atomic numbers could readily be identified and plotted against their position in the periodic table of elements. Verification of this, demonstrated conclusively that a sequential plot could identify elements based upon their atomic number rather than their atomic weight as Mendeleev had originally presented.

Before proceeding farther, some thought must be given to how the electron(s) is (are) configured around the nucleus of the atom specifically in an expositional sense in that more than one have been found and, of equal importance, why the electron(s) remains in orbit to render the total elemental charge neutral and how the electrons interrelate to other elemental proximities. In accordance with the theory of Quantum Mechanics, atoms and electrons can exist in certain allowable states, whereby each state is defined by a specific energy level. When an electron changes its normal energy state to another allowable one, that is to say an orbital level exchange, it is the result of either an absorption or emission of a sufficient amount of quantum energy to bring that change about. Accordingly, each atom possesses a quantum number associated with those individual electrons comprising it and which define that energy of the specific electron according to its summation relationship insofar as the orbital speed required to maintain its distance from the nucleus and its energy relationship to the positive field of the protons.

The periodic table references each element in accordance with its atomic number and offers a direct relationship to the number of electrons in orbit about the nucleus of that particular element. As the electron count is advanced by one, its place within the periodic table is incremented by that amount or to the next higher number.

Each shell within an orbital level can contain a maximum of two electrons, and the description of this structure from an orbital viewpoint is called the quantum (a basic and fundamental unit of energy) number and referred to as *m, l, n, o,* and *p* (fig. 2.6). These identifiers refer to the quantum numbers for *m* as 1, 2, 3 . . . and so forth. The identity of *l* refers to the azimuthal quantum number with designates 0, 1, 2, 3 to n-1. The third quantum reference, *m,* is to the orbital magneticquantum value with its designate 0 or any positive or negative integer up to and including *l.* Measurements, that is to say, the absorption of X-rays, showed that those atoms considered to be light or possess loosely attached electrons will vibrate and thus emit an electromagnetic frequency (which will be discussed in detail later) that can be measured. For these cases, *n* was equal to one-half the atomic weight (1969:60). Not only did this type of investigative work determine a more precise structure of the atom, but it also revealed information relative to the electron envelope and spacial distances between atomic centers within a molecule. This is most important when considering that

Element		K	L		M			N		Ground Term	Ionization Potential (in electron volts)
		1, 0 1s	2, 0 2s	2, 1 2p	3, 0 3s	3, 1 3p	3, 2 3d	4, 0 4s	4, 1 4p		
H	1	1	—	—	—	—	—	—	—	$^2S_{1/2}$	13·59
He	2	2	—	—	—	—	—	—	—	$^1S_0$	24·56
Li	3	2	1	—	—	—	—	—	—	$^2S_{1/2}$	5·40
Be	4	2	2	—	—	—	—	—	—	$^1S_0$	9·32
B	5	2	2	1	—	—	—	—	—	$^2P_{1/2}$	8·28
C	6	2	2	2	—	—	—	—	—	$^3P_0$	11·27
N	7	2	2	3	—	—	—	—	—	$^4S_{3/2}$	14·55
O	8	2	2	4	—	—	—	—	—	$^3P_2$	13·62
F	9	2	2	5	—	—	—	—	—	$^2P_{3/2}$	17·43
Ne	10	2	2	6	—	—	—	—	—	$^1S_0$	21·56
Na	11	Neon Configuration			1	—	—	—	—	$^2S_{1/2}$	5·14
Mg	12				2	—	—	—	—	$^1S_0$	7·64
Al	13				2	1	—	—	—	$^2P_{1/2}$	5·97
Si	14				2	2	—	—	—	$^3P_0$	8·15
P	15				2	3	—	—	—	$^4S_{3/2}$	10·9
S	16				2	4	—	—	—	$^3P_2$	10·36
Cl	17				2	5	—	—	—	$^2P_{3/2}$	12·90
A	18				2	6	—	—	—	$^1S_0$	15·76
K	19	Argon Configuration					—	1	—	$^2S_{1/2}$	4·34
Ca	20						—	2	—	$^1S_0$	6·11
Sc	21						1	2	—	$^2D_{3/2}$	6·7
Ti	22						2	2	—	$^3F_2$	6·84
V	23						3	2	—	$^4F_{3/2}$	6·71
Cr	24						5	1	—	$^7S_3$	6·74
Mn	25						5	2	—	$^6S_{5/2}$	7·43
Fe	26						6	2	—	$^5D_4$	7·83
Co	27						7	2	—	$^4F_{9/2}$	7·84
Ni	28						8	2	—	$^3F_4$	7·63
Cu	29						10	1	—	$^2S_{1/2}$	7·72
Zn	30						10	2	—	$^1S_0$	9·39
Ga	31						10	2	1	$^2P_{1/2}$	5·97
Ge	32						10	2	2	$^3P_0$	8·13
As	33						10	2	3	$^4S_{3/2}$	10·5
Se	34						10	2	4	$^3P_2$	9·73
Br	35						10	2	5	$^2P_{3/2}$	11·76
Kr	36						10	2	6	$^1S_0$	14·00

Fig. 2.6
Quantum Number Reference
Atomic Electron Distribution
(Atomic Physics:1969-181)

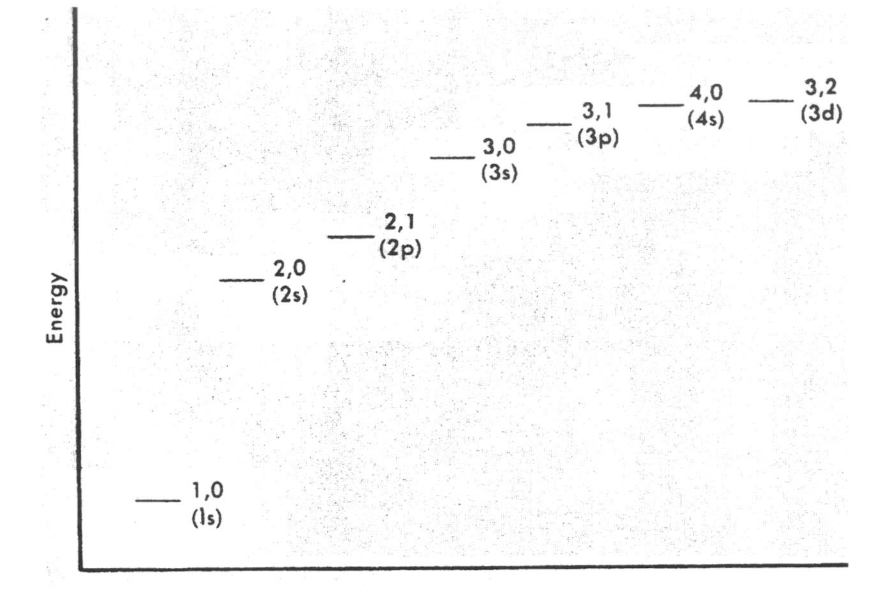

Fig. 2.7

Energy Possibilities for Subshell Configurations
(Chemical Principles:1966-173)

reactions between atoms are determined by the number of outer electrons in the cloud surrounding the nucleus and not the mass of the nucleus itself. Here we have the first look at what is responsible for reactions between the elements themselves from the atomic number of the element and possible predictabilities to sustained chemical properties.

Predictable theories can be made using studies of the electron cloud rather than the nucleus itself. To continue this atomic study, a closer examination of the electron cloud would be required to satisfy several questions regarding it. Can this cloud alter its shape so as not to be spherical about the nucleus and can electrons be taken or added to a nucleus in ways that would cause binding forces to be altered or identified? Also, can one element be changed into another by the addition or removal of electrons?

The Elusive Electron

We have established at this time that discoveries conducted by Rutherford, Moseley, and Mendeleev substantiated various electron existences within multiple and layered orbital levels. The single electron within the hydrogen atom as an example, or any atom with a multiple electron structure, is attracted to the proton in such manner that the closer the electron is to the proton, the stronger the attractive force (1987:46).

The configuration representing the various levels within an atom as determined by the various quantum numbers representing its state: i.e., those components that encompass an electron's energy and motion dynamic probability are four per electron, whereby only two of the numbers, *n* and *l,* will be discussed here, as they indicate the most relevant factors.

To begin with, the number *n* is a reference to the level or shell (1966:172) in which the electrons belong, and it should be noted here that the level capacities are somewhat limited and will increase as the number *n* increases. In addition, the higher or farther the electron is from the nucleus, the higher its energy value would be. The subshell quantities are fixed and determined by the quantum number *l* and may contain the following electron quantities. $l = 0$ may contain two electrons, $l = 1$ may contain six, $l = 2$ can contain 10 electrons, et cetera (fig. 2.7).

From the dynamic standpoint of electron motion between shells and among levels as theorized, some energy must be imposed upon the electron and of sufficient value to provide the impetus required for a transition to occur. Also, should an electron assume a higher level, than its quantum numbers would also reflect that particular level quantum identification. To further illustrate what has been covered regarding any electron configurations and the transition dynamics surrounding the nucleus, an additional understanding of their level relationship and orbital behavior must be undertaken to clarify that picture.

To satisfy the requirement of electron configuration and potential in the terms of their motion and charge, that interaction will be called electronic energy. The change in energy the electron will either absorb or emit to satisfy the conclusion to the changed state or level is called a quantum of energy, and it is that energy that will identify an electron characteristic. The energy involved will identify particular frequencies of radiation that can be useful in displaying specific elemental involvement.

There is evidence that not all electrons revolve about the nucleus in a

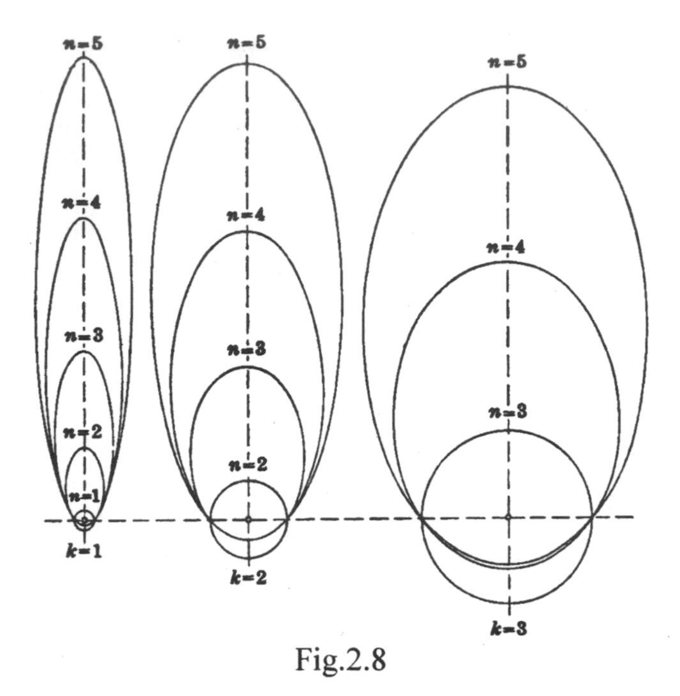

Fig.2.8

Bohr-Sommerfeld Electron Orbital Configurations
(Atomic Spectra and Atomic Structure:1944-18)

circular orbit. In some cases, the electron moves in an elliptical orbit (fig. 2.8) with its axis rotating slowly about the center of gravity and not about the nucleus itself (Herzburg, 1944:16). This is due to the mass change of the electron as it alters its proximity to the nucleus in the revolution. Also, in reference to figure 2.1 concerning the electron cloud, the electron does not occupy all the relative space at the same time. The electron represents a point charge whereby the density of the cloud at any given reference point may indicate the probability of finding an electron there. As to the mention of a circular orbit, figure 2.9 demonstrates that electron motion about the field lines of the nucleus is syncrotronic in its dynamic. The reasoning behind the spiral is due to the inability of the electron to escape the field line generated by subatomic relationships. As the electron lacks ability to escape the field of energy it circles and it has a velocity component, a spiral is manifested.

In any consideration regarding a philosophy behind elemental changes, there is a requirement to provide the concepts that express radiative properties as a consequence of identifiable behavior associated with nuclear orbital displacements (fig. 2.10). As energy is applied to a nuclear structure (that in most cases is a quanta of specific wave mechanics), some part of the total energy spectrum may stimulate an electron into a proximal change. It is this part of the orbital characteristic of the electron and what happens to it that is most important in any consideration regarding interactions with other nuclear structures leading to potentials that alter basic elemental configurations to other than original. For an expositional rational of any nuclear reaction, some fundamental aspects behind the electron's reply to stimuli will greatly assist in our understanding regarding this nuclear dynamic.

Since elements are specified by the number of electrons as designated by the atomic mass, considerations first would be to identify where they would be in an example structure such as hydrogen. Figure 2.11 demonstrates that probabilities certainly exist giving an approximation of not only where they might be but also their relationship to any spherical cloud formation. The figure shows that the probabilities of finding an electron near the midradii point is greater than at a distance from the center and, as the more solid lines indicate, here we also have the greatest probability of electron density.

So OK, we can estimate for a quantum number n the probabilities of finding an electron within a specified cloud density, yet to complete the understanding of the atom additional information is needed. What are the transitional states of the electron and why do transitions occur? Last, how do they alter a nuclear state and create new structures?

First in the identification of electron states, definitions must solidify our fundamental understanding of level energies with respect to electron behavior. In the previously discussed atomic structure, orbital states for the electrons were identified to be in association with certain energies with respect to the positive charge within and about the nucleus. To this extent investigations show that the number of electrons, up to five, would naturally form in a single ring (Longmans, 1924:293). If six electrons were associated with the nuclear structure, an unstable occurrence would be in effect. To stabilize the structure, five electrons would relocate in a second ring and one (the sixth) would be found to occupy the closest level or ring about the nucleus (fig. 2.12). With reference to this table, it can be seen that the

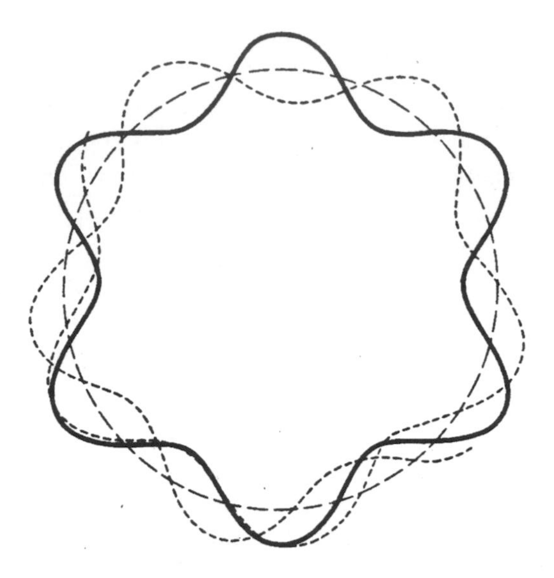

Fig. 2.9
De Broglie Waves for Circular Electron Dynamic
(Atomic Spectra and Atomic Structure:1944-31)

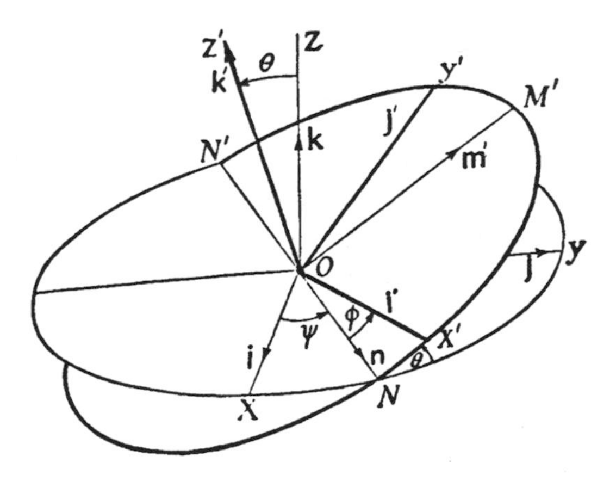

Fig. 2.10
Basic Orbital Structure Angular Relationship
(Theoretical Physics:1956-139)

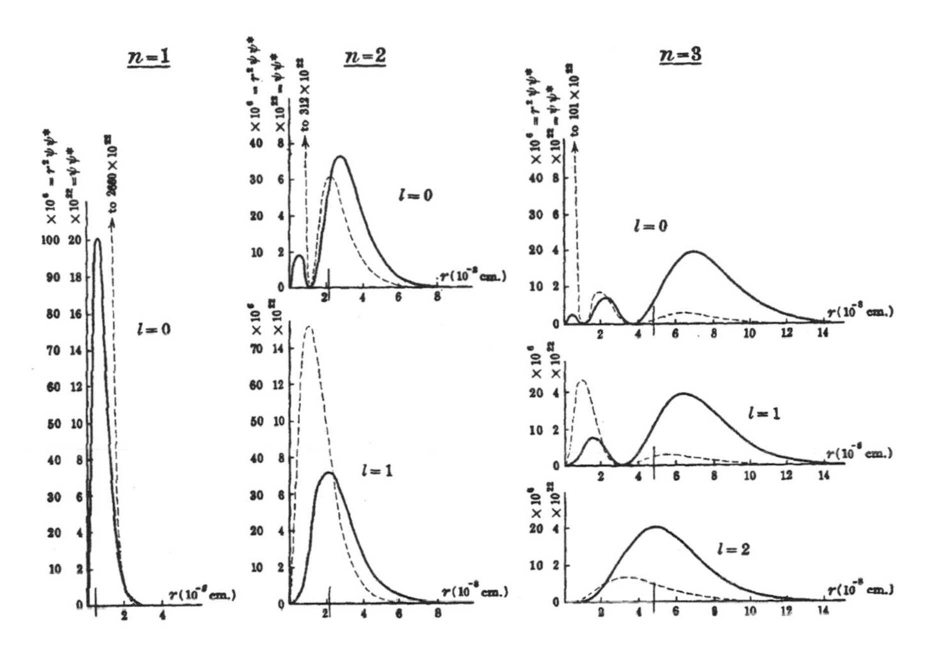

Fig. 2.11

Probability of Density Distribution (Hydrogen)
(Atomic Spectra and Atomic Structure:1944-43)

atomic structure will reorient itself to stabilization depending upon the electron quantities with respect to the periodic elements.

Since electrons will radiate (covered shortly), a stable situation will only occur if the electron is not radiating energy. Also orbital stability will occur if the electron had a quantum relationship between its orbital frequency in revolution and the energy of the electron. If this was satisfied, then and only then would the electron remain in a permanent orbit (Longmans, 1924).

A Outer ring	1	2	3	4	5													
B Outer ring	5	6	7	8	8	8	9	10	10	10	11							
Inner ring	1	1	1	1	2	3	3	3	4	5	5							
C Outer ring	11	11	11	12	12	12	13	13	13	13	14	14	15	15				
2nd ring	5	6	7	7	8	8	8	9	10	10	10	10	10	11				
1st ring	1	1	1	1	2	3	3	3	3	4	4	5	5	5				
D Outer ring	15	15	15	16	16	16	16	16	16	16	16	17	17	17	17	17	17	17
3rd ring	11	11	11	11	12	12	12	12	12	13	13	13	13	13	14	14	15	15
2nd ring	5	6	7	7	7	7	8	8	8	8	9	9	10	10	10	10	10	11
1st ring	1	1	1	1	1	1	1	2	3	3	3	3	3	4	4	5	5	5
												a	*b*	*c*	*d*	*e*	*f*	*g*
E Outer ring	17	18	18	18														
4th ring	15	15	15	15														
3rd ring	11	11	11	11		etc.												
2nd ring	5	5	6	7														
1st ring	1	1	1	1														
	h	*i*																

Fig. 2.12

Electron Orbital Configurations

(Ions, Electrons and Ionizing Radiation:1924-294)

Nuclear Radiation

In 1665 Sir Isaac Newton held a prism to the sunlight in such a way as to produce a multiple assortment of colors on an opposite wall (fig. 2.13). With this observation, he wrote that the sunlight passing through the prism exhibited a spectrum of diverse colors. He used the word *spectrum,* noting a divergence in colors and, that there was an implication of specific colors. Later, in 1802, a more refined examination of the colors was undertaken by William Wollaston to enhance further a definition of their specifics. He noted seven distinct variations comprising two variations in red, three in the variation of yellow-green, and two representing a blue-violet color (Hamilton, 2001:102). This was significant in the fundamental understanding that there was something within and from the sunlight that generated a variety of colors one could not perceive in a normal way. Also, that it was very similar to the rainbow effect normally viewed after a rain storm, only artificially generated by establishing a light-to-particle velocity relationship.

Fig. 2.13
Spectral Representation of Sunlight Components
(*Modern Earth Science*: 1983-31)

The total light from the sun apparently included all the various colors that were exhibited by the prism. The entire spectrum of the colors viewed was within the light called visible light. This is the light that our eyes perceive and translate into colors and images. As the sunlight translates through the dimensional material of the prism, it is bent or changes its direction to an angular relationship with respect to the incoming light and is offset. At this junction however slight as it may be, the individual particles associated with the total sunlight composite are moving in different velocities.

Because of this extra (angular) difference or "bending," the sunlight's composites at their respective velocities will vary to exit the prism at different times. The difference in the velocities will then establish the difference in speed, as colors. Red would be the slower-velocity particles, and the white, violet, et cetera, would be representative of the higher speeding particles. This was an important verification into evidence that suggested that the sunlight we see is a composite of many particle speeds. It is also suggestive that an atom may consist of many representations concerning its nuclear dynamic and that perhaps it, too, contained information that could be uncovered and made visible for recognition. It should be noted here that prism colors represent only a portion of the sunlight's energy composites and that there are many high-velocity particles within the sunlight that we will never see without special detection equipment.

As for our discussion with respect to the exposition of an atom's structure through particle detection, the emission of atomic energy and the dynamics that accompany the process are identifiable. This was considered, and a defining structure of the atom was proposed in 1913 by Niels Bohr, who was one of Rutherford's students. This proposal was titled "On the Constitution of Atoms and Molecules" and printed in volume 26 of the *Philosophical Magazine.* In this distinctive reading, the explanation of the hydrogen atom was defined by Bohr using the concept of radiating frequencies between an unbound and bound electron, whereby all the energy of the electron is emitted as radiation from the procession of an initial unbound to bound state (1990:107). In this case, the unbound state is represented when the quantum number n reaches infinity. The total energy of the system W and the frequency of orbital motion f equal zero. The system's energy represented by W is a mathematical equation where the electron's mass m is multiplied by the square of its orbital velocity, minus the electrons magnitude e and the nuclear charge Q (coulomb), divided by the orbital radius r.

Although the Bohr structure will not explain emission intensity and is incomplete for multiple electron configurations and chemical bonds, it was helpful in showing an existence of a stationary state in that the most stable state with the most energy is one with the quantum n equal to 1. In addition, it demonstrated the relationship between electron energy and orbital quanta and, most important, that emission will occur from an electron as it alters its orbital position about the nucleus. In this manner, Bohr was able to show that as an electron passes from one orbit to another orbit, energy (quanta) is released (emitted or radiated) that has a certain frequency. Therefore, if an electron with a certain energy level attained in the second orbit about a nucleus has the energy level it possessed while in the first orbit subtracted from it, the remaining frequency is the radiated amount. It is then concluded that electron transitions between orbital states can be calculated to represent frequencies (1924:129).

The radiated energy is the nuclear (electron) radiation that can be measured and defined by electron quantity and orbital configuration. In reference to figure 2.14, it can be seen that the larger the quantum number n, or orbital levels present within an atomic structure, the higher the radiation frequency will be. Certainly where there are atomic structures that are configured in such a way as to possess many electrons and many orbits this concludes that there will always be transitions. Therefore, there will always be a form of radiated energies that will cover a wide frequency range.

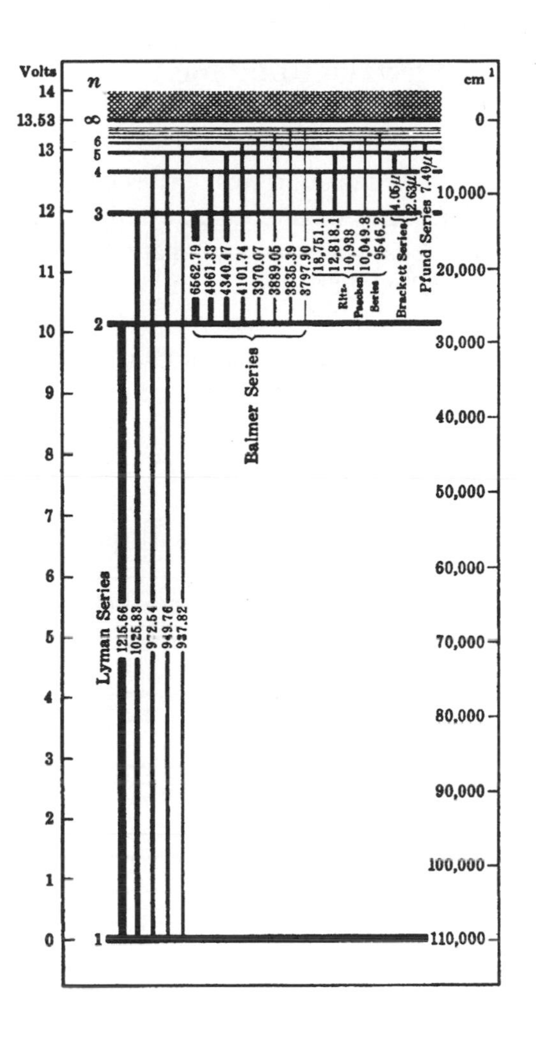

Fig. 2.14

Energy Level of the Hydrogen Atom
(Atomic Spectra and Atomic Structure:1944-24)

Electron Rotations

The lowest range is the Lyman Series (fig. 2.14). To explain that there are many different and diverse radiation frequencies being emitted, an investigation of known frequencies should be addressed. Using the quantum numbers representing *n,* where *n* is equal to 1, that is called ultraviolet. *N* equal to 2 represents the Balmer Series or visible light (that which we can optically see); where *n* is equal to 3 is referred to as the Paschen Series or infrared, and if *n* is equal to 4, that represents the Brakett Series, also infrared. Beyond those frequencies and even higher in *n* number is the Pfund Series.

In reference to figure 2.14, again the *n* number is illustrated by the horizontal solid black lines, the electrical energy level designated in cm (to the minus 1) is referenced on the right scale and represents the spectral wavelength. It can be seen here that the closer the electron is to the nucleus, i.e., *n* or ground state, a proximital change in orbital characteristics to release an electron from the protonic balanced charge or ionization potential can be related to wavelength energies about 110,000 cm (-1) and may be identified by the appropriate angstrom number. The thickness of the line represents the relative intensity.

An electron's stimulation or excitation potential at the ground state of the hydrogen atom will not occur below a radiated frequency of 1215.66 in angstroms (covered later) because of its proximity within an energy level that is closest to the nucleus. At the quantum level of *n* equal to infinity, separation between the proton and electron will occur with the least amount of electron voltage applied.

Excitation within an atom will in general occur when the farthest or outermost electron (least energy bound) is stimulated to move into a next higher or outer level. Other excitations of electrons may take place if two or more outer electrons are moved to a higher level or one of any inner electrons are raised accordingly within the lower levels to an outer one (1944:127). Having stated this, there are certain laws that must be observed with regard to when an atom can reside within various states, and that is covered under the Pauli Exclusion Principle (1924). In simple terms, the Pauli principle is based upon electron magnetic spin or a quantum number *ms*, within an orbital level.

According to W. Pauli, no two electrons may have the same set of quantum numbers (1990:83). This is important when there are atoms with multiple electrons in residence about the nucleus. Also, if a condition exists whereby there is more than one electron or electrons present, they cannot all be in the lowest energy state but must be forced to occupy more elevated energy "shells" of higher and progressive energy. In view of this thinking, the farthest electrons will feel a shadowing effect from the closer electrons nearer the nucleus. This effect is partially responsible for the farthest electron's ability to be removed with the least amount of energy.

So far, only general terms have been applied to an electron's quantum condition(s); a more explicit understanding of emitted and absorbing states should be carefully examined. This is important in any atomic understanding because it leads to a more secure and fundamental understanding of conditions and expectations that may be applied to a planetary atmosphere of the nature that surrounds us all.

Let us turn our attention once again to figure 2.8, where closer investigation will yield some additional reference points for analysis. In figure 2.8, a variance in orbital ellipse possibilities as the principle quantum number k is represented by the function of the sum of $k + n$ and is so illustrated. In addition, in this pictorial with each k possibility are associated energy levels with the quantum number n from 1 to 5. These figures represent in quantum mechanics the relationship between an electron in an orbit and the distances that require an electron at the various levels of n to rotate about the nucleus within such an orbit.

Since an electron in orbit will rotate about the nucleus with a certain periodicity, as such, a certain rotation time is defined. In addition, because a certain level of energy is required to define that period of rotation, it must come from the characteristics of its orbit. Therefore, if an orbit becomes larger, then the period of rotation will also increase and with that period of rotation, an electron will emit a frequency that corresponds to the type of orbit that it occupies. In an elliptical orbit, as the electron velocity increases near the nucleus, that difference is also emitted as a frequency called a harmonic frequency (Bohm, 1979:38). If the orbit is circular, then no velocity shift would be detected and that particular rotation frequency emitted would be considered the fundamental frequency. Because atoms of matter are varied in configuration, the orbiting electrons will emit frequencies based upon that particular arrangement and always for that particular atomic structure. These frequencies are measurable in many forms; one particular method of viewing an atomic structure's identity is by displaying the emitted frequencies in a spectral line.

The identifying frequencies of an orbital electron determine a period

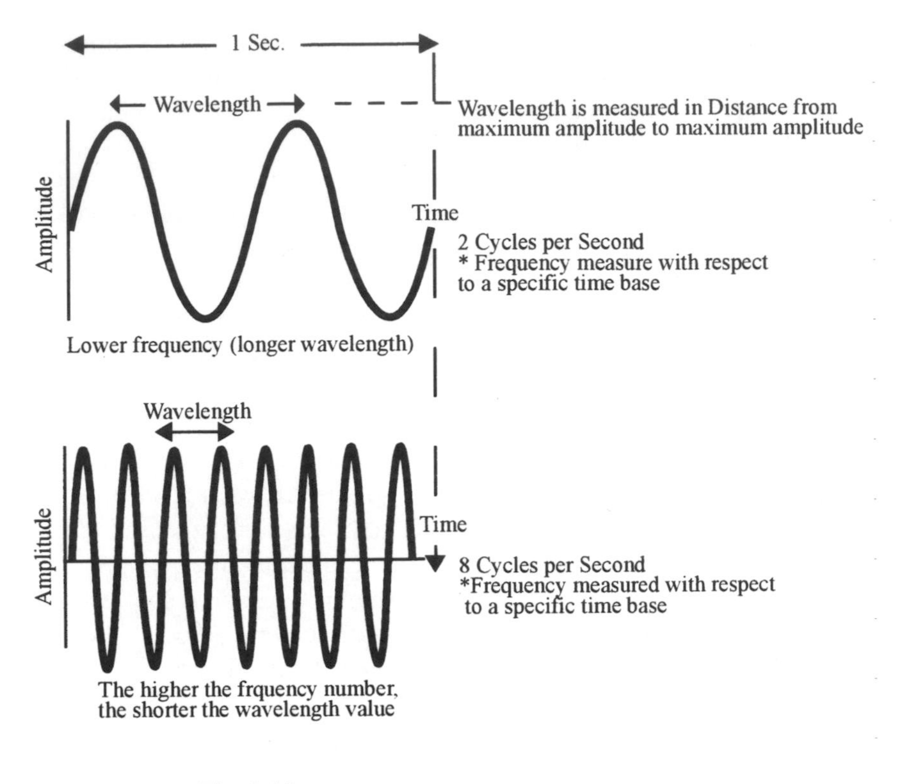

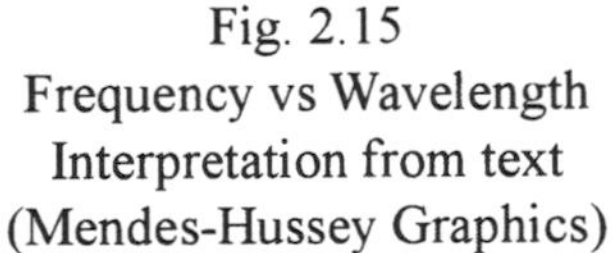

Fig. 2.15
Frequency vs Wavelength
Interpretation from text
(Mendes-Hussey Graphics)

of rotation, and since the energy is radiated outward there must be some additional component of that frequency, which also is of a determinate value. That value is the distance in which the photonic emission has traveled outward, as the result of an orbital rotation within a period of time, identified as the wavelength of that frequency.

The distance is measured in centimeters, and the acceptable representation is the Greek letter *lambda.* With reference to figure 2.8, it can be understood that in larger orbits if the electron has increased velocity, it will start and finish at a certain orbital point quicker; therefore, the rotation repetitions are sooner and thus shorten the corresponding wavelength distance (fig. 2.15). This is in agreement with a higher frequency in cycles per second termed *hertz* and the corresponding shorter distance between them measured in angstroms, whereby one angstrom is equal in distance to one hundred millionth of a centimeter.

Electron Emission

In figure 2.15 a distinct change between a positive transition and negative transition can be seen. This is termed an *oscillation* and as such is created by an oscillator. In quantum terms, an electron is also recognized as a rotating oscillator, and for consistent physical terms we will apply that here.

Energies, that is potentials in existence that will cause an object (rotational oscillator) to alter its established behavior by overcoming inertia, are established by the difference between one atomic energy level (orbit) and a transition by the oscillator from that energy level to another, referred to as quantum states. The relevance in this exposition will manifest itself when future discussions concerning atmospheric dynamics are covered. Energy in its relationship to heat and light via rotational oscillators will become a consequential part of this learning.

In 1900 Max Planck suggested that atoms and the smaller particles associated with those structures may only posses a certain and definite amount of energy (1966:168). Most energy that was readily defined could be found in heat, and with this as a reference, Planck formed a hypothesis that considered the emission and absorption of energy as it relates to matter, stating that the radiation concerned with substances does so in finite amounts (quanta). In addition, the radiation concerned is also consistant with certain standards of energy quanta multiplied by its frequency or repetitive cycles within a specified amount of time (*hv*) (1969:82).

Because the Earth is considered a body of matter, according to Planck it will be a radiator and absorber of energies. Therefore, concerning ourselves with the radiator at this time, Planck's theories should be understood to build the quantum foundation necessary to further identify how the rotating oscillator will behave along with its associated energy expectations under certain conditions. By definition (Planck), a black body is a perfect absorber of energy and will also emit the maximum amount of isotropic energy. That is, its intensity will be independent of the emitting photonic direction (1992:95) at a specific temperature.

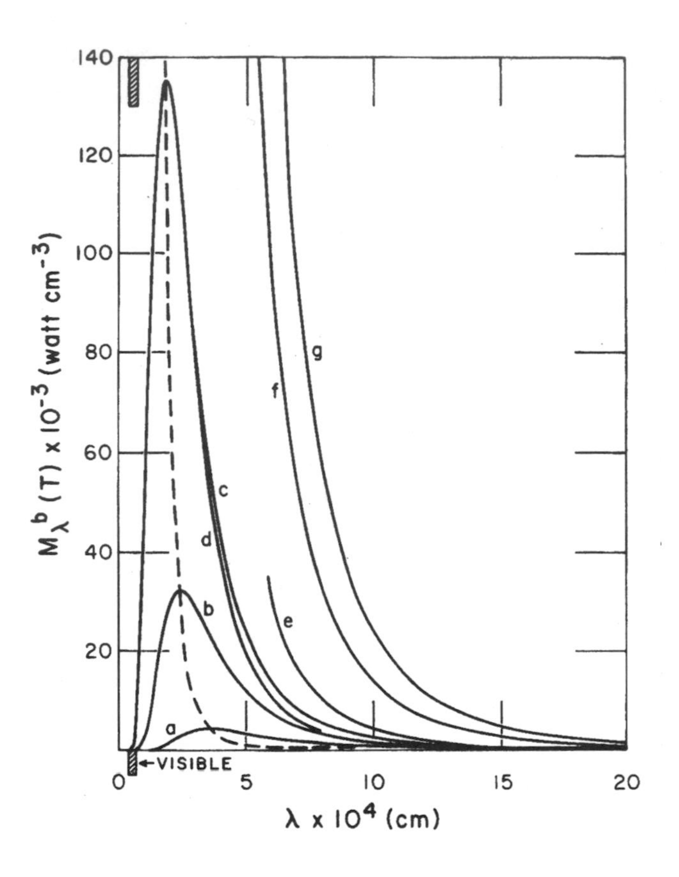

Fig. 2.16

Plank's Blackbody Relationship
(Encyclopedia Of Physics:1990-105)

The emitting and absorption of energy quanta can be qualified above its electron configuration. A gaseous state where the electrons are loosely packed within a finite area will emit quanta semiindependent of the internal atomic structures because of the lack of compact interaction. In an area where the atoms are greater in density, as in a solid or liquid, when the atomic structures are excited into an emission of photons the interactivity between the structures themselves will emit additional quanta. This addi-

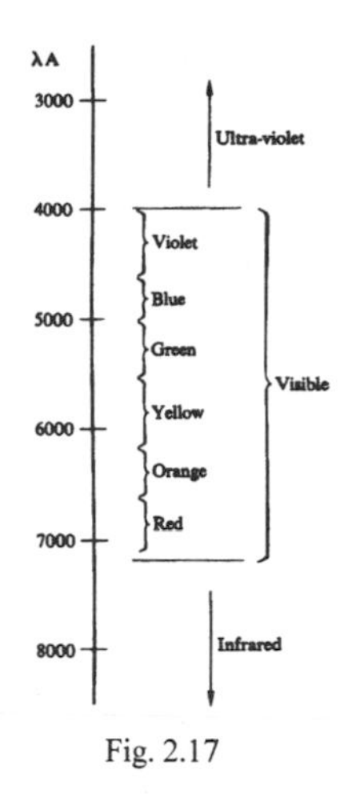

Fig. 2.17

Optical Spectrum vs Angstrum (Lambda)
(Introduction to Electronics:1967-236)

tional emission associated with structure interactivity is called thermal radiation. It should be noted here that this type of radiation is identified with that of the infrared frequency range and associated with heat sources (1990:104).

Figure 2.16 explains this blackbody relationship in greater depth between wavelength and frequency. In this figure, note that the intensity of the wavelength (right side scale) is measured against the wavelength spectrum in centimeters with the visible portion located at the bottom left side. The greatest intensity, a temperature of 800 Kelvin in wavelength, occurs about 2 x 10 to the fourth power in centimeters. The line represented by (*g*), or at a temperature of 10,000K, possesses an intensity of emission much greater around 7 x 10 to the fourth power in centimeters. Accordingly, in figure 2.16 most of the intensities lie between reference lines at (a) to (g), and represent those temperatures between 800K and 5,000K that are indicative of the infrared spectral regions.

With this information, several points become clear from the text. First, the frequency of the wavelength (v) in cycles per second is equal to the velocity in meters per second (c) divided by the wavelength (*lambda*) in meters as defined by angstroms. Figure 2.17 will illustrate this conversion and diagram the angstrom value to the optical spectrum for further enhancement. It may be seen that the greater the distance between the repetitive cycles through larger angstrom numbers, the lower the frequency will become, and a progression from ultraviolet to infrared can be accomplished.

In addition, to make the conversion into wavelength more accurate and reflective of the photonic energy, a constant was derived by Planck and assigned the physics identification *h,* at a value of 6.624 x 10 to the -27 erg sec. Since photonic energy is indicative of wave emission, the total energy (*E*) of the photon is then the result of multiplication of both frequency (*v*) and the Planck constant (*h*), or E = *hv*. The processes for exciting atoms into emission of light vary. Although principle emission is temperature radiation, other sources may be electroluminescence and chemoluminescence. With temperature radiation, emission of photonic energy is by way of the kinetic energy of internal atomic (atom or molecule) collisions and is generally accomplished through high temperatures. In emission by luminescence, electrons are excited into an ion situation whereby the electron has been raised into an energy state that completely separates it from nucleus attraction and is free to escape and collide with another atom or molecule. In chemolumenescence, the energy in an atomic structure is liberated through a chemical process by reaction. In photoilumenescence, however, as the word implies, the photonic light is emitted when an electron absorbs energy.

Radiation emission intensity can now be better understood and identified as the number of transitions taking place in one secondd multiplied by the photonic energy designated *E.* Within the factor *E* is a representation of two important items. One is the number of atoms in the initial state of the level *n,* and the other is the number of transitions per second. In measurement of absorption intensity, the change is from atoms to molecules in the initial state (*m*), the radiation density of the particular frequency(*v'*), and a new term (*B*), which is indicative of the transition probability for absorption (1944:152). In addition, transitions by one electron will be far more intense than if two or a multiple of electrons jump at the same time.

Just as an excited electron will jump from one quantum level to a higher one, if an electron is in a higher energy state level then a prediction that the electron in a higher quantum level will radiate energy when it drops to a state of lower energy is also possible. In any consideration regarding additional energies that an oscillator may be subject to absorbing, that oscillator must be subjected to a source, or in physics terms a "field," of energy. It is this field that will be discussed and analyzed further.

The Electric Field

Absorption of energy (quanta) by an electron is possible through a relationship with several forces within a specified space, which are acting upon it. The force structure, if in harmony with the oscillator, will generate excitement that eventually causes the transference of that oscillator from one quantum level to a higher one. As with any influencing energy, there will always be the factors of intensity, type of radiation force, i.e., wavelength, frequency, and of course time. All should be addressed in any investigative work that define parameters required for electron stimulation.

The term *field* in this analysis refers to any area that contains electric, i.e., quanta, forces that will either repel (cause to move away from) or be attracted to (cause to move closer) other similar electric forces in a specified vicinity. Let me state here that I consider an electron a potential force by itself because it reacts to other electrons and displays quanta variations depending upon outside stimuli. As a unit unto itself, it possess quanta energy that can be directed through motion and either be influenced or influence other electrons in proximity. Proton quanta can also be considered a form of force, because it is a balance to the quanta possessed by an electron and will allow a stable atomic structure. The proton, if on its own, would certainly be an influence to a single electron, as we have seen, and, as investigated, is opposite in nature and characteristic to it and equal with respect to their mutual quantum differences.

Possession of a potential that influences other opposite and similar quanta is called a charge, and, for concurrence, the dictionary lists this quanta or protonic and electronic potentials as a charge and is defined by the physics symbol (*e*). The potential of this unit of charge is 1.602 x 10 to the minus-nineteenth power and is defined as coulomb. Even though the proton has the same unit of coulomb charge, its mass is 1,836 times as great and is about the same size as an electron. The charge these atomic structures maintain can neither be created nor destroyed (Branson, 1966:25), although they may be individually separated from a complete atomic structure that would contain both in a balanced form.

When there exist multiple charges in a specified area whereby they will react individually to the proximity of others contained within the same confines, then an electric field is defined and in existence (Lemon, 1934:217). Accordingly, the distance that both charges would need to be separated by to produce an influence on each other of opposite but equal

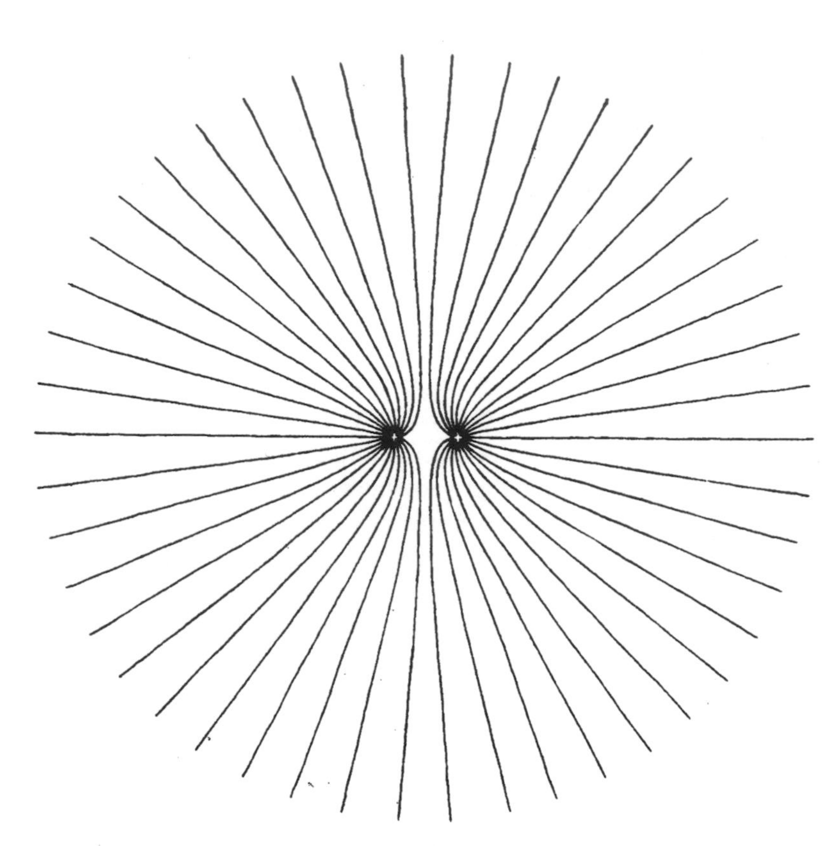

FIG. 2.18

Equal Potiential Quanta in Proximity (repell)

(From: Of The: Earth, Spheres and Consequences:2001-94)

potential is one centimeter in any direction. The image of an electron or proton as having a coulomb charge of one dyne is then identified by a spherical configuration about the structure with the point of the coulomb charge potential located at the center. Since like charges repel and unlike charges attract, the figure of like charges between two protons or electrons in proximity to another considering the spheroid dimensions would look very similar to figure 2.18.

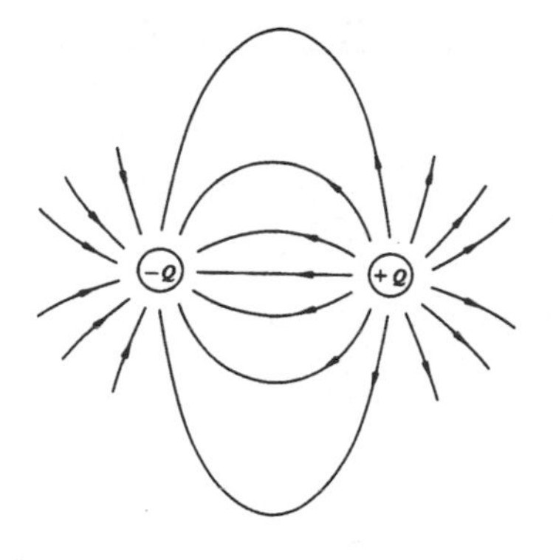

FIG. 2.19

Unlike Charge Field Lines

(Introduction To Electronics:1967-31)

To further understand, the force influence upon a grouping or between charges is stated as the force exerting influence at a specific point is an electric field at that point (Branson, 1966:29).

With regards to the field of protons and electrons, lines of force are represented as inward flowing (fig. 2.19); this is because a positive charge would be attracted to a negative one. A presupposition of intensity becomes important at this stage due to the logical assumption that there exists some appreciable distance between the quantums in their relationship to the individual representative atom.

In the case of the atom that has undergone a quantum stability change by the presence of an electric field (*E*) (fig. 2.20), the positively charged nucleus (*K*) moves away from the electron's(s') established center of gravity (*S*). This destabilization of the previous center (*S*) in relationship to its motion along the field line (*E*) will cause a self-correcting action to take place. The action by the atom will of course attempt to realign its new electron's(s') center of mass (*K*) to the field line. The newly formed line (*J*), still being ninety degrees perpendicular to the new center of mass's rotation about the field line (*E*), now begins a procession that is related to the offset magnitude (see the dashed line within the smaller orbital circle) to the new orbit about the field line.

The rotation velocity of this new offset arm's (*J*) orbit about the field line *E* is dependent upon field line strength and will be constant to the orig-

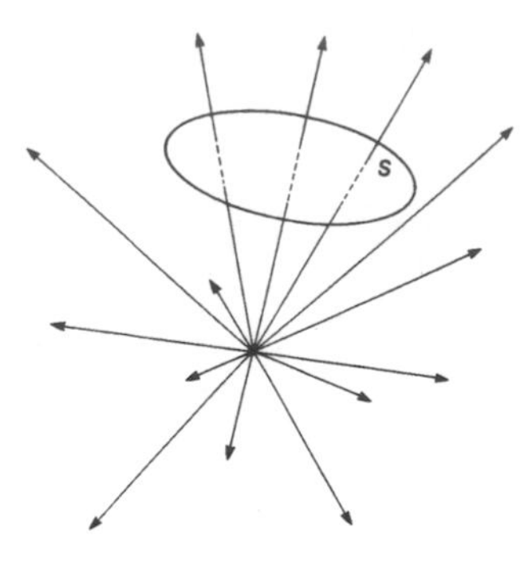

Fig. 2.20

Point Charge Field Line Emanation

(The New Physics:1989-468)

inal nucleus's position and distance shown as (*M*). Figure 2.21 establishes this cause and effect with regards to atomic activity. In structures where there are more atoms and their quantum levels are several, the magnitude of shift (small dashed line) would depend upon the relative distances from electron neighbors. When multiples of n levels exist, accordingly the distances between electron shells and occupants become smaller. In the case of higher levels of energy states and associated transitions, when the shift does take place within them their transitional velocity will affect any emitted wavelength (refer to fig. 2.17).

Stated previously, an attractive force is present between two unlike charges at one centimeter's distance; viewed as a singular reference, this force is at any point in a radial direction from the center where the point charge is located (fig. 2.20). Should an atom pass or become close enough in proximity to the field of a quantum charge to be affected by it, as demonstrated by the positive nucleus being offset in some distance, the electron cloud surrounding the nucleus will distort because of that partial movement in separation between the positive nucleus negative charges' mass center contained within (1991:271). When the distribution of electrons within the orbital cloud is induced to be disturbed in this manner, the result is a polarization of that atomic structure (refer to the plus and minus designations in fig. 2.21). The positively charged nucleus will move along the line of force separating it in distance from the center of electron gravity creating a nonzero electrical dipole moment (1944:115).

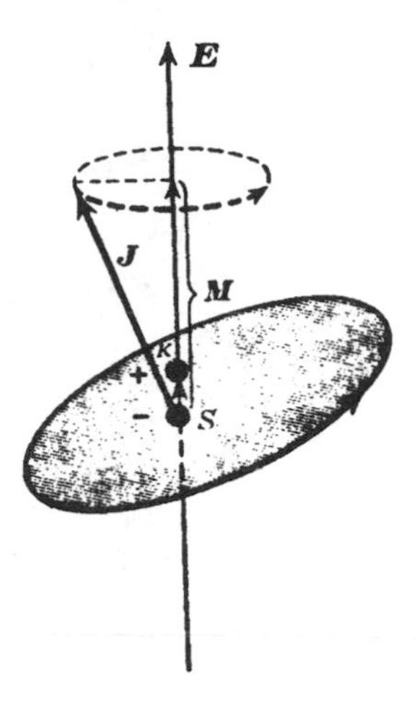

Fig. 2.21

Electric Di-Pole Moment
(Atomic Spectra and Atomic Structure:1944-115)

The change from a monopole to a dipole configuration is most important when factors such as photonic emission, ion contiguousness, and multiple atomic structures are in proximity to one another. To place all that in the context of this study, remember that the Earth's protection is primarily based upon an electromagnetic field that protects us from harmful solar particle emission. Also, since atmospheric constituents are comprised of nitrogen and oxygen atoms that are in free spacial surroundings, add to that the additional unifications of the hydrogen atom with resident oxygen atoms in the production of the water molecule.

With these factors ever present around us, further investigation of the exposure to particles beyond the monopolar when in reference to electric fields and quanta must be included to exemplify the total atomic structure around us. Only then will this provide a solid foundation for explorations into particle behavior, which we depend upon so much.

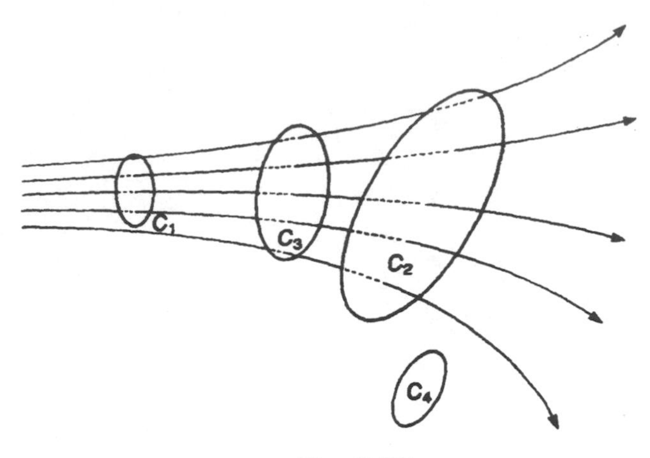

Fig. 2.22

Magnetic Lines of Flux defined thru an Area

(The New Physics:1989-468)

Electromagnetics

Proximity of a quantum of an energy field that is stable and that of an atom will cause the atom to distort. This will occur without the quantum charge or atom being in motion; that is, neither is moving; they are just near enough to cause an atomic dipole. If the electrical charge is moving, than a magnetic field is in existence about and in the vicinity of it (Lemon, 1934:233). Essentially, a magnetic monopole has never been determined to exist anywhere as a point charge; even in the magnetic point of a bar magnetic, the lines of force continue along and through it in the direction of the negative end (1989:468). In the field of magnetic lines as defined as the "field," the flux of the magnetic field is the number of the lines of force that pass through a specific area encircled about those specific lines of force (fig. 2.22).

In the consideration of the lines themselves regarding field generation and the atom, with reference to figure 2.21 again, the dashed line represented as a radius generated by the moment orbit will pass directly through the field line *E* and extend an equal distance away from it in a straight and opposite line. There is still a precession; however, the point of orbit is at the end of the extension. I believe that this extra magnitude is the difference between the electric field and the magnetic field and can be differentiated from an electric field by the expression of vectors (1951:74). The similarities between both electric and electromagnetic will be the Planck constant and frequency; also, both lead to the conclusion that a physical force is involved.

It is important to recognize that the invisible magnetic and atomic electrical moments are a relationship to the fact that the nucleus is a charge potential and, because of its spinning, is analogous to that of the spin of an extranucleus electron, also in motion. Because of that motion and the fact that it has mass, it also will develop an electromagnetic momentum additional to the entire atomic structure and which would be smaller than that of the electron (1944:186).

Electromagnetic fields generated by charges in motion can be applied directly to the environmental situations around us. It is precisely because the magnetic field will induce behavior in atomic structures, termed a *magnetic moment,* that we should be quite aware of its presence and longevity. In previous discussions regarding fields, we have taken the atom and subjected its structure both in totality and individual nucleons to quantum forces. As the atom is considered to be matter by itself, a closer look into the field–matter relationship can now be undertaken to further define field existences and their generation through forced (induced) nuclear alterations.

How can we develop a magnetic field through the rearrangement of matter? How is it possible to see the invisible fields and determine the patterns generated by nuclear forces in motion? Last, why do we care about such matters when searching for the definition of atmosphere?

Man was not always aware that there was a force that not only was invisible to him but also was capable of having physical influences discernible by mankind. As early as A.D. 121, the Chinese, through the suspension of certain types of metal rods, noticed they would always point in a specific direction (Branson, 1966:132). Small pieces of a metal substance, a substance that was not yet identified as being metal, much less recognized as of polarable for that matter, was encountered near the Greek city of Magnesia in Asia Minor. Also, those inhabitants, too, found that not only would these pieces have an influence on each other, but they also would have an invisible influence on other objects. The fact that an influential forces exerted itself in a variety of intensities and always at a geographic terrestrial north location eventually gave rise to the name pole, both north and south. The strength of the attraction or repulsion of the poles is determined by several factors. The strength is inversely proportional to the distance between the different charge locations such that the farther the distance, the smaller the influence. Another factor governing influence strength is the product of the pole strengths, which is proportional to the force (influence) of its attractive or repulsive charge. Two men named Karl Friedrich Gauss and Wilhelm Eduard Weber are responsible for the mathematical calculations and discovery of the aforementioned magnetic properties named for them.

Because a magnetic field will be created when a charge is in motion, it is most logical to reason that an electron can be excited by a quantum of energy to become removed from the quantum levels *n* about the nucleus easier than a proton may be excited to removal from the nucleus. With this reflection, then the direction of motion would be in the direction of the removed electron(s)' progression. It is this progression to a quantum charge of attraction that is electrically considered the flow or current. However, in understanding that a positive and a negative nucleon of charge will have motion differences, today the conventional wisdom defines current to be the direction of the positive charge (Branson, 1966:43). Having said that, the atomic flow is the electron moving toward the positive. Here is motion of charge; therefore, by definition, a magnetic field will be in proximity to the charge itself. If a positive pole and negative pole were separated by distance, a charge potential would exist around and between them to establish a force of influence demonstrated in figure 2.19. If a piece of substance was polarized, that is, made to have more protons than electrons or more electrons than protons at specific ends, one of the ends would react accordingly to the magnetic field or influence so diagrammed, if placed near or within it.

In utilizing the property of electron flow or current, a polarity of either positive influence or negative influence can be caused in metal so that it is in effect a dipole having a magnetic force. This is in effect causing the atoms within the metal structure to align themselves in a particular arrangement in that the positive nuclei will distance themselves from the electron cloud. Obviously, the metal must be able to internally be atomically aligned, and this is called magnetic susceptibility. The amount of time that the metal will remain in a dipolar state can be either permanent, as in the case of a hard steel, or not, as in the case of soft iron.

Because the positively charged nucleus will move in the direction of a

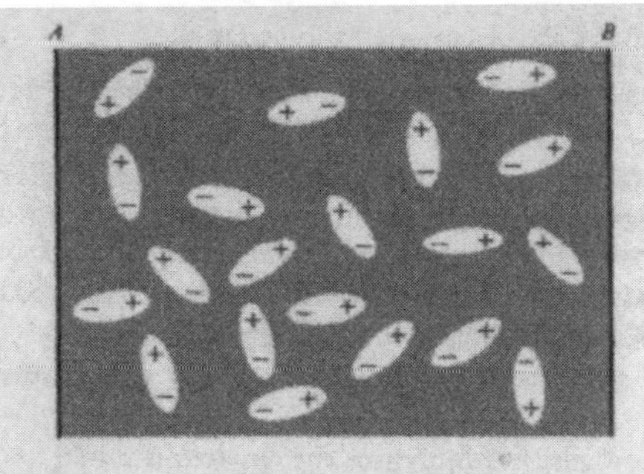

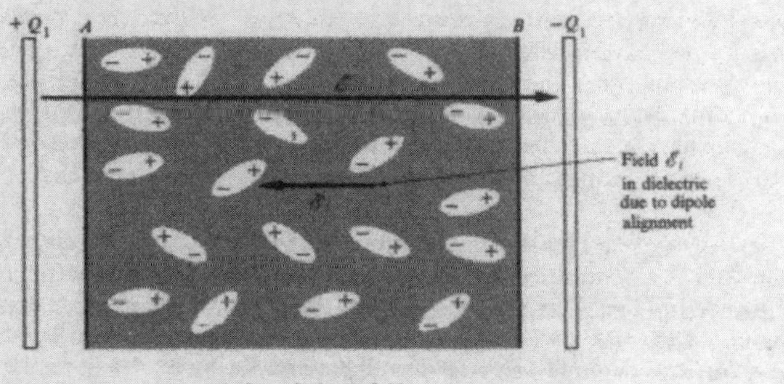

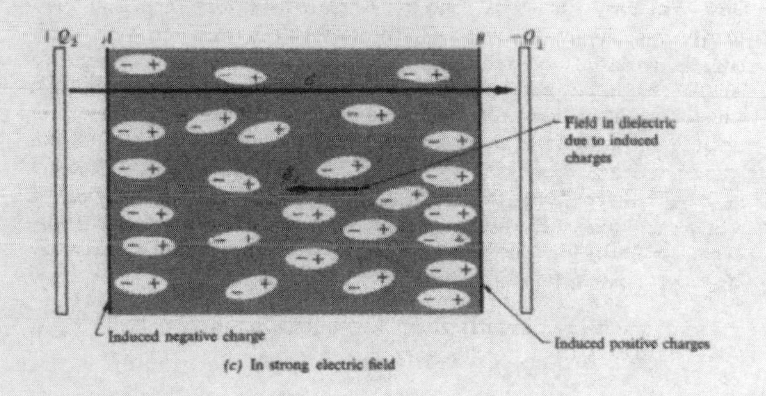

Fig. 2.23
Atomic Alignment in a Field Line
(*Introduction to Electronics*: 1969-110)

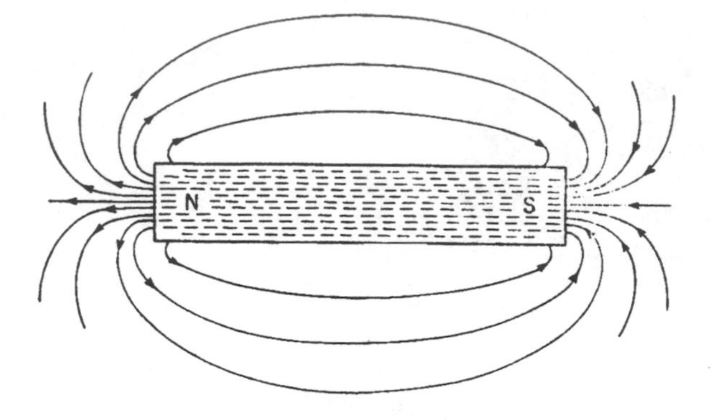

Fig.2.24
Dipolar Configuration with Magnetic Field
(Principles of Electricity:1961-20)

field line and away from the electrons, this motion will enhance the alignment of the structure within the field line. Figure 2.23 demonstrates this nicely, and it may be seen that the positive charges have moved into an arrangement whereby they all "face," or are orientated along, the field line accordingly, dependent upon the strength of the field line to which they are in proximity. The longevity of their orientation depends upon the material that contains these atoms and their ability to be susceptible to orientation and remain in that position once the field line has been removed or continues to remain in existence. It is precisely because the positive nuclei in the atoms of the substance in their dipolar configuration orientate themselves accordingly that the induced positive end of a dipole, or in this case the positively charged end of metal, will "point" toward the north. We can draw a conclusion that our Earth must then have some form of field lines in proximity so that a dipolar structure will be induced or made to cause an orientation. If this is indeed the situation, then figure 2.24 would be very close to the actual terrestrial representation. As we will study later, this Earth's electromagnetic field, or called geofield, is in fact created through the internal heat and metal core. Because a field of electromagnetic energy affects atoms to a reorientative state, this effect can be important with combinations as well. Atomic structures of nitrogen, hydrogen, oxygen, and even solar particles all exist around us, and because of this, they are all subject to the geofield force.

Although the study of the atom and its relationship to electromagnetic and quantum forces can now be better understood, the fact still remains that we are surrounded by atoms of many kinds with their own unique structures. This surrounding has great meaning, well beyond the fundamental ontologies that were postulated by the atomists.

3
Discernible Behaviors

Atmospheric Invisibilities

Webster's Comprehensive Dictionary of the English Language (1998) defines atmosphere as "a mass or body of gases that surround the earth." Although the dictionary also uses the definition "a pervasive element or influence," I truly think that both meanings are applicable in this instance.

In both cases, the fundamental aspect that defines the atmosphere would be in the influence it has upon us, in such a way as to cause notice to be brought to it. The influence from its components would then bring attention if our well-being was threatened by them. Can the components of this surrounding gaseous body affect more than a physical relationship to us? If a particular reason could be found for us to want a mental state of pleasantness, could that state be induced or influenced by this envelope of surrounding gas? The answer is yes. Many reasons exist every day that point to this influence. A sunny day at the beach would require the components of this surrounding gas to be without moisture. That would be a pleasant situation for both our mental and physical state. A wedding, barbecue, or sports gathering would also require the gases' cooperation to provide the "influence" of surrounding us without moisture, which then would translate into the enhancement of our mental condition.

To be sure, all is not always pleasant with this gaseous surrounding. Other factors must be considered that would balance this pleasantness with regard to our state of mind with the components of the gaseous surrounding. Considerations of motion also require us to realize an influence upon us. References to these gases' exhibiting dynamic motion such as that of a tornado, hurricane, or cyclonic dynamic will not bring pleasant feelings but instead bring sadness and loss. Again these situations focus attention on the gas envelope surrounding the Earth, and because of a realization

Fig. 3.1
Potential Atmospheric Indicators
(Lutgens, Tarbuck: 201-272)

that we are being influenced to such a significant level, a priority consistent with the word *subjugation* starts to be established.

As an example of servitude to the envelope of gas, some tangible evidence should be introduced giving reasons for the consideration of "subjugation" and associative priorities when attempting to define influential relationships pertaining to mankind's existence and dependence. Since our

Fig. 3.2
Apparent Absence of Atmosphere
(Space view of Mt. Tambora, Indonesia)

existence is based upon the ability to survive, the understanding of that capacity to extinguish through inherent properties must become a focus. This exposition of mankind's susceptibilities to the surrounding gaseous envelope will identify those influential properties so important to the atmosphere that we must know.

We have determined that there is truly an existence of an atomic presence in this sphere; what it is and where it is must then be explored. Further, is the atomic component all there is to our sphere or is there something else in partnership with the atomic structures that also prevents our demise? Viewed from a position above the Earth, the only perception we can see regarding an existence of the atmosphere around and over the planetary surface appears to be in the consistence of cotton-puff balls exhibiting what could be represented as some form of density (fig. 3.1). In this image, they are seen to be hovering or in some way suspended over the landscape. If indeed they are suspended, then questions arise as to the form of suspension. Are they suspended in an envelope that surrounds them much like some milky substance suspended in a clear liquid? Or are these densities floating on something that appears to be invisible to our eyes?

Further inspection of this image identifies a shadow underneath. This would indicate that whatever the substance is does not pass light through it and that the light's absence is cast as a shadow, also that the surrounding light about the cottonlike substance does not identify any trace of matter that could be seen to occupy the space between the cottonlike substance and the Earth's terrestrial surface.

In other analysis regarding Images that have been taken to identify parts of the Earth's topographic nature, some cases will also not identify any existence of a substance that we would consider from perception, as an atmosphere between the Earth's surface and the orbital camera. The white imagery represents ice and snow accumulation (fig. 3.2). So, as one considers these points or specifically the lack of evidence as to any atmosphere or surrounding gaseous envelope, an ability to locate what we can and can't see must be undertaken if any analysis of this "atmosphere" issue is to have credibility in its identification.

One of the first steps undertaken to indicate a location of the atmosphere is to inspect the image 3.1 closer and compare that with other images that have been taken at an altitude sufficiently above the Earth that would produce some visible evidence of an atmospheric substance. In figure 3.1, the substance seen appears sporadic, and if one would consider it to be atmospheric in nature, then how is it possible that life may be sustained with such few examples? In order to ascertain if these white substances are the same over the entire terrestrial surface, then a more distant view of the surface area must be done to determine this. In figure 3.3, the majority of the Earth's surface from a distance of some length exposes a more defined pattern of what appears to be a similar substance, although its motion would indicate that it flows much as a river of liquid would. Again, although the patterning is different and the substance does look more fluid, can this be the atmosphere that we search and is the substance seen both from low and high altitude the same? Reasoning at this point would tend to steer thinking into a confined process of determining if the atmosphere is

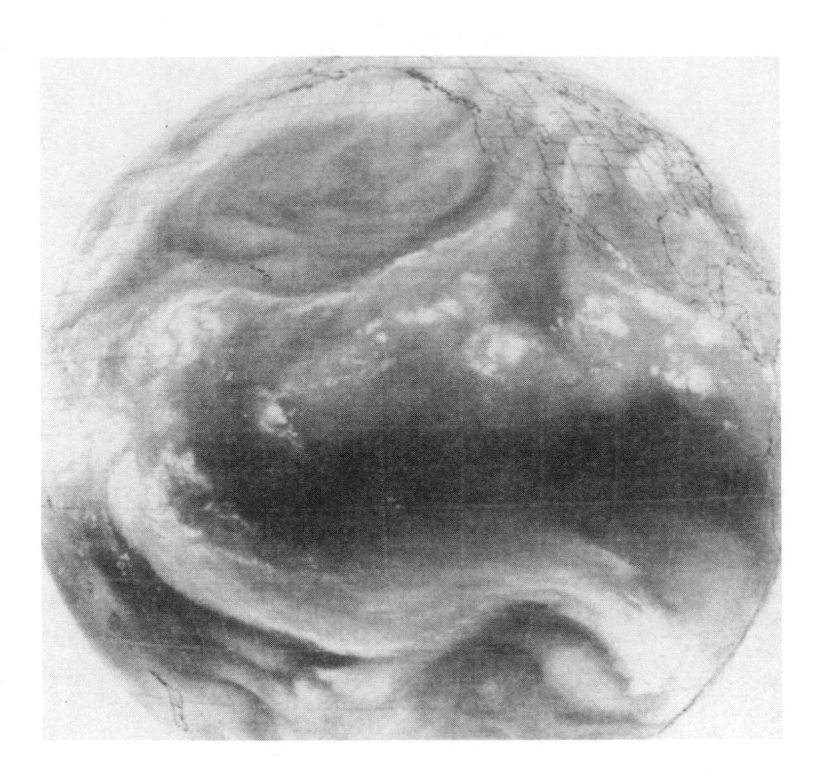

Fig. 3.3
Earth View of an Atmospheric Substance
(GOES Western View [1982] NOAA)

only of the substances given without further thought to another possibility, or if the possibility of the substances seen is not the atmosphere, then we still lack evidence of where this atmosphere is and the environment of its visible properties. In addition, some thought as to the height of this substance may suggest the possibility that wherever we are searching for may be found at extreme distances from our terrestrial surface.

Wherever this "atmosphere" is may require multiple properties that work together at several heights to create the totality of its definition. Previously some discussion of the definition of the "atmosphere" exposed considerable thinking by the early Greeks. They put forth an exquisite philosophy introducing the concept that because we cannot see what is around us that does not mean that something is not there. They presupposed the word *atom,* and since from that time we have expanded that thinking substantially into further explorations of their definition, which is that the atom is indeed small enough to be undetected by our eye and not further divisible into any smaller component or segment. Interestingly enough, because it was small enough that we could not see what was considered to be defined as matter, precisely because it was matter, it therefore had to have some mass. From that reasoning was developed the rationale that it must have weight. By establishing that consideration we may successfully determine where the weight of unseen matter may reside and to what extent we may further define that existence.

The Elastic Fluid

Remember the Atomists? They felt that there was something around us although they at the time could not really give the "something" any real identity. As stated earlier, we breathe it, whatever it is. Birds and aircraft are able to move upward because of the physical force generated by their motion through promoting a lifting effect. We can see many behaviors in diverse physical objects such as trees, waves, and various items that seemingly either are being moved into motion or have a floating appearance, such as the cottonlike substances that produce water and hide the sun around and above us. Truly there must be something there around us, but we cannot see whatever it is. So is there anything at all?

The first century A.D. was a milestone in attempting to determine if anything at all even existed that would bring an explanation to the invisibility of our surroundings. An open container was placed upside down in water by a Greek engineer of that time named Hero; he observed that the water level would not increase any higher in the container until the invisible air that was trapped was let out. With this experiment, he concluded that air must possess some form of substance to have prevented additional amounts of water entering its space until whatever the invisible substance was left (1999:48). This of course did not mean that the invisibility of the substance had weight, but that there was indeed something present that acted as a force pushing against the water. Here was the first physical observation of the definition of opposing forces in action; we today call this pressure. Pressure did nothing to define in that time period what the substance was, nor did Hero recognize the pressure for what it was; however, there was something to be said for his experiment.

With the fact that the Atomists mentioned "atoma" and that these particles were around us gave light to other thinking that perhaps some form of science could be linked to the identification of these structures through their combinations. The first attempts were from the ancient world, where the Chinese and Indian cultures attempted to reform certain metals into gold. We call this pseudoscience alchemy. Later, around the twelfth century, Hellenistic Europeans tried to turn lead into gold; even though their attempts failed, within that effort was created the real science of chemistry.

Although alchemy was eventually relegated to the ash bin of discreditation, its research continued into reactions and constituent expositions. Aristotle was perhaps one of the earliest Greek philosophers to recognize that the invisibility of a substance around us was important in our lives. He did not, however, share the Atomists' view of its structure, perhaps because he preceded them by so many years, and in the two hundred or so years additional concepts would have an influencing edge. Aristotle gave the alchemists fuel in his creation doctrine that through heat and dryness properties of substances may change, i.e., lead into gold. Since the alchemists were of a certain spirit in all things, they applied that concept in that the spirit could be brought back into something else; therefore, the lead to gold theories had some validity for them. Aristotle also believed that "air" was an element, that is, it was all there was and not divisible into any smaller parts. This air relationship was applied to all things by him and therefore by his doctrine was in all things in existence. It was part of his peripatetic proposition that all things were composed of earth, fire, air, and water. He was adamant in the use of the word *element,* even if he could not take that meaning any further, and as a consequence the name of atom was not used in the science world without bringing rejection and for a while became lost (McDonnell, 1991:64).

Later during the period of the 1640s an Italian physicist, Pierre Gassendi (1592–1655), was very faithful to the revivification of the atomic structure and its meaning to science. He suggested that matter was created by God and in that creation matter was composed of solid small and impen-

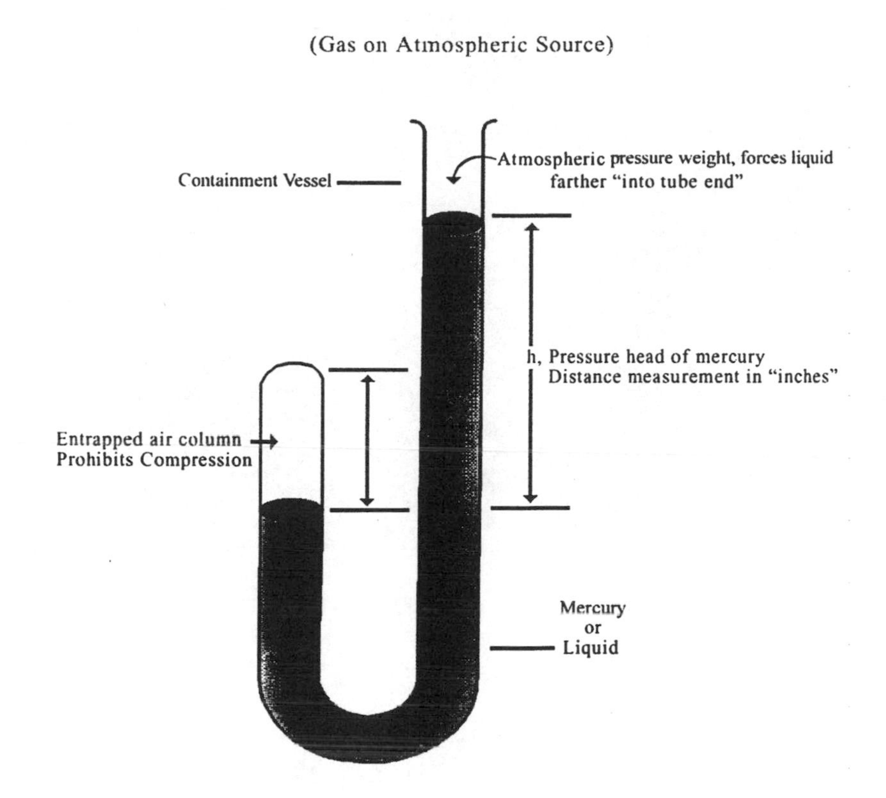

Fig. 3.4
Interpretation of Boyle's Experiment
(Mendes-Hussey Graphics)

ctrable particles, a theory that Sir Isaac Newton (1642–1727) also supported (1966:30).

In 1748, a Belgian chemist by the name of Johannes Baptista van Helmont (1579–1644) was experimenting with, at that time, aeriform liquids. His work published by his son in 1648, Ortus Medicinae, contained his new term for the invisible composition, which he called gas. There was the first name for invisible matter.

Aristotle maintained through his writings a loyal faction by the scientific communities for many years; however, in 1661, through the printing and presentation of a work called *The Skeptical Chemist,* Robert Boyle buried Aristotle's element philosophy and concepts by opening the door to

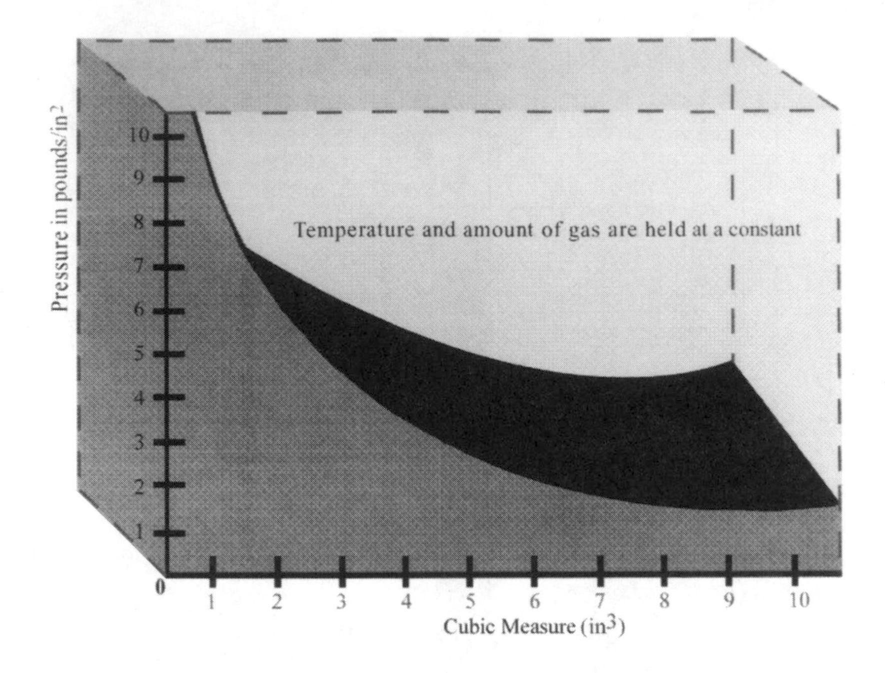

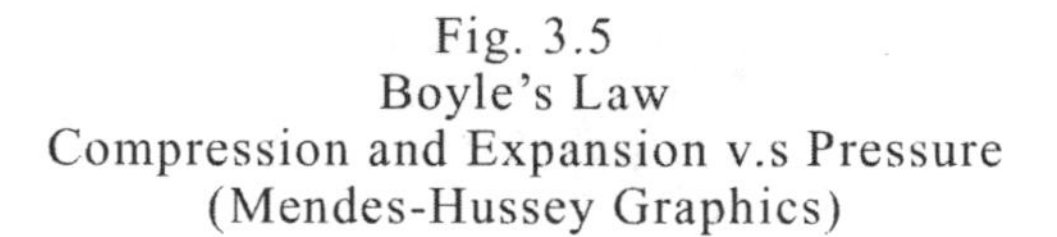

Fig. 3.5
Boyle's Law
Compression and Expansion v.s Pressure
(Mendes-Hussey Graphics)

what is now the foundation of modern chemistry. Robert Boyle was born in 1627 as the seventh son of the first Earl of Cork and died in 1691. He was educated at Eaton and became recognized as a natural philosopher, conducting his experiments once he acquired laboratory premises at Oxford. It was here most of his research into the field of matter was conducted in or about 1654–68. During 1662, this research identified and concluded what we now have and use as a standard today, that which is known as Boyle's Gas Law. Boyle's experiment in the deductive reasoning behind the fundamental principles of the newly-termed gas as it would apply to the invisible substance around us was identified through a unique but simple experiment (fig. 3.4). Even if at that time the composition of air was not yet determined, there was evidence that the invisibility of the substance could be betrayed by compression.

As other gas properties were not evident by their lack of discovery,

Boyle could only use the air around and available to him. This was certainly sufficient to produce the data that was most convincing and applicable to other gas discoveries. Within the simple glass tube was liquid mercury. When the air pressure was increased at the open tube end and the tube and its environment were at relatively constant temperature with respect to the mercury column head, the mercury would rise into the closed end. The trapped air pocket would decrease in volume inversely proportional with the applied pressure. As long as the temperature and gas amount is maintained at a constant, figure 3.5 will illustrate Boyle's Gas Law.

Here was the first explanation of the invisibility of matter and its relationship to a surrounding environment, in that an existence had been proved however its composite structure remained undefined. Boyle, it seemed, was satisfied with this discovery and, although influencing others such as Sir Issac Newton and John Dalton, did not distinguish whether the invisibility was atomic or corpuscular.

Isaac Newton did not subscribe to the entire Atomist doctrine. He did recognize that small particles were in existence but allowed additional reasoning that departed from the experimental and verifiable. He felt that atoms could be divided by the act of god (McDonnell, 1991:80). Also, with regards to gas in its composition as an elastic fluid, Newton felt that any fluidic elasticity could be explained by mathematics regarding volume and particle distances. These were assumptions on his part using geometric equations, and although they agreed with Boyle's Gas Law, Newton's position on the assumptions he had to make in his hypothesis were quite numerous. He assumed a volumetric symmetrical spacial distribution and when under the influence of compression the particle quantity in numbers and structure did not change, also that repulsion was mutual in a gas (elastic fluid). Size was not considered as well in his mathematical explanations. Although Sir Isaac Newton presented a good case for elastic fluids, his hypothesis was considered to be only a possible explanation and was not supported by actual evidence.

Atmospheric Identity

The later seventeenth century was quite prolific as far as atmospheric constituents regarding atomic structures were concerned, the beginning of which was through experiments in the area of combustion. Until the following individuals presented their discoveries about the primary gases in the 1770s, scholars in academia were positioned to accept the fact that combustion was sustained because of a presence within the breathable air called phlogiston. This was a Greek assumption originally, that in essence there was a substance (*philogistos*) incorporated within our breathable environment that was the primary cause for the flammability of objects.

In 1772, this assumed component of the atmosphere was considered a reasonable path of explanation by two men, Joseph Black (1728–99) and Daniel Rutherford (1749–1819). Both of these two men believed phlogiston a reason for combustion to occur in free air, although Mr. Rutherford is credited with one of the actual atmospheric components, that which he called phlogisticated air and we now call nitrogen.

Joseph Black was one of Rutherford's teachers at that time in the university, during 1772 in Glasdow, and it was Dr. Black who began to experiment with what would result in the eventual discovery of carbon dioxide. It was through Dr. Black's classroom problem that one of his students at that time, Mr. Rutherford, discovered the gas nitrogen.

By studying the properties of carbon dioxide through experimentation, Dr. Black discovered that in a confined area a burning candle would eventually die out. The first reason was that the burning candle produced carbon dioxide, so through a chemical process he removed that amount as well. When he tried to endure any combustion in the air that was left, that process would not support any sustained ignition. He reasoned that the exhaustion of the component of phlogiston must have been the reason for that situation. This was a reasonable answer to the extinguishable characteristic, for it was also his philosophy and belief on the matter. What this remaining air sample was became the problem that he turned over to one of his students to resolve. That student was Daniel Rutherford.

Daniel Rutherford attempted to resolve this problem of just what the remaining parcel of air was. Born in Edinburgh, he also became a scientist and attended the University of Edinburgh. It seems that Rutherford verified the air that was left over in Dr. Black's experiment to the extent that his explanation was along the very same reasoning as Dr. Black. Because candles could not be reignited and mice died, both the burning candles and respiratory functions of the mice gave off phlogiston to such an extent that the air surrounding these experimental components had caused it to become saturated with the phlogiston substance. To this end, I am not quite sure that Mr. Rutherford discovered the real gas identity as the extinguishable component within the air, which would have been the answer to Dr. Black's problem, but simply reconfirmed its existence through the verification of Dr. Black's experimental result. As it came about, Mr. Rutherford agreed that the remaining air must have also been phlogiston. Today, however, we now know that the remaining air that caused such extinguishing properties was nitrogen, and Daniel Rutherford is credited with that discovery.

Henry Cavendish (1731–1810) presented in 1785 the idea that the Glasgow experiment had indeed isolated a gas that existed within the atmosphere and the properties of that gas were consistent with Rutherford's thesis. The first part of the atmospheric was in place. The final exposition of the remaining component was not far away; however, the replacement of the phlogiston theory had not yet been presented.

Joseph Priestley (1733–1804), was the first to discover through experimentation the existence of a new and additional atom within the immediate air. He was born in Yorkshire and eventually gained acceptance as an English ministerial theologian in Leeds and later a scientist, and his efforts are now considered the cornerstone of the chemical realm that we breathe today. Early in his ministry, which was near a brewery, he was intrigued by the gas created from the fermentation process. With this expanding curiosity he endeavored to conduct experiments on what was then carbon dioxide as well as two other gases from the only two available and known at that time, the air around him and hydrogen. In 1774, through laboratory efforts, he accomplished what was eventually considered to be the most famous discovery of that time, although he did not recognize the resultant atomic structure for what it eventually became known as.

Priestley heated mercury oxide with the sun's rays, using a large lens to establish a focal point. During this process, a colorless gas was given off, which, when objects were ignited in it, caused a much brighter light effect than normal. This gas substance he termed *dephlogisticated air,* and in turn he mentioned this experiment to another scientist at that time by the name of Antoine Lavoisier.

Antoine Lavoisier (1743–94), a French chemist, enhanced the work done by Joseph Priestley to identify and further understand the gas that had been discovered. Because Lavoisier was a pragmatist of sorts, he held the belief, as did the Atomists, that there was a cause and effect regarding Priestley's experimental resultant gas. Lavoisier's fundamental position was that atoms or small particles of matter were responsible for the experiments' outcome. Since Lavoisier's laboratory contained chemical apparatus that was more advanced and precise than previous scientists, he was able to gather data in a more conclusive manner. His conclusion from observation and combustive experimentation was that objects burned not because of the previously considered component of phlogiston but because of the mixture of the objects matter composition with the new invisible gas that Priestley had uncovered. With this revelation, the phlogiston component of air was finally discredited and replaced with the atom of what Lavoisier termed *oxygen.* Because of this experimental work and diligence of exposition, the Aristotelian view that the invisible air about us was an element became relegated to history and an incorrect assumption. In 1789, Lavoisier published his synthesis of chemical knowledge at that time through the work *Trait'e ele'mentaire de chimie,* which many in the chemical field consider the first chemical textbook.

In this publication, Lavoisier presents a systematic layout of the chemical representations as we know them today and removed the belief in only the four elements of creation that had been held since Aristotle. Lavoisier had concluded by the time of his death in 1794 that the Aristotelian element of air was not an element at all but a mixture of gas that comprised the atoms of oxygen and nitrogen (McDonnell, 1991:92).

4

The Cohesion of Energy

Elemental Considerations

The two principle gases of density that constitute what was then considered breathable air at the beginning of the eighteenth century consisted of nitrogen and oxygen. Further thinking presented itself with respect to composition of atomic structures in their participation of atmospheric quantities, and because of this, questions regarding atom weight and percentage of volume, as well as the current discoveries in oxygen and nitrogen, were in the constant presence of additional nondiscovered atom expectations.

Fortunately, a German scientist by the name of Jeremias Benjamin Richter (1762–1807) was also thinking about atoms and the relationships they may have to one another, not necessarily within the atmosphere but in a more liquid form such as in reactions between acids. His experiments consisted of attempts to render neutral-specific bases through the interactivity of various mixtures. This was defining in the way chemical mixtures reacted to form specific results. The mixtures had to be exact. Acceptable results from the experimentative process dictated that it was not a close measurement. This thinking was in line with Proust's in that there was a specificity in the atomic world regarding mixture and that some indefinite amount could not be satisfactory when involved within a chemical range measured in exactitudes.

With this information and experimental results, Richter concluded that there was something called equivalent weight that would only be acceptable in the expectant result of a chemical mixture's reaction. This work was published in 1792 and considered a cornerstone with respect to the Law of Equivalent Proportions. He presented that two equal amounts of different elements will combine with a third. An example of this is: one gram of hydrogen will combine with eight grams of oxygen to form water.

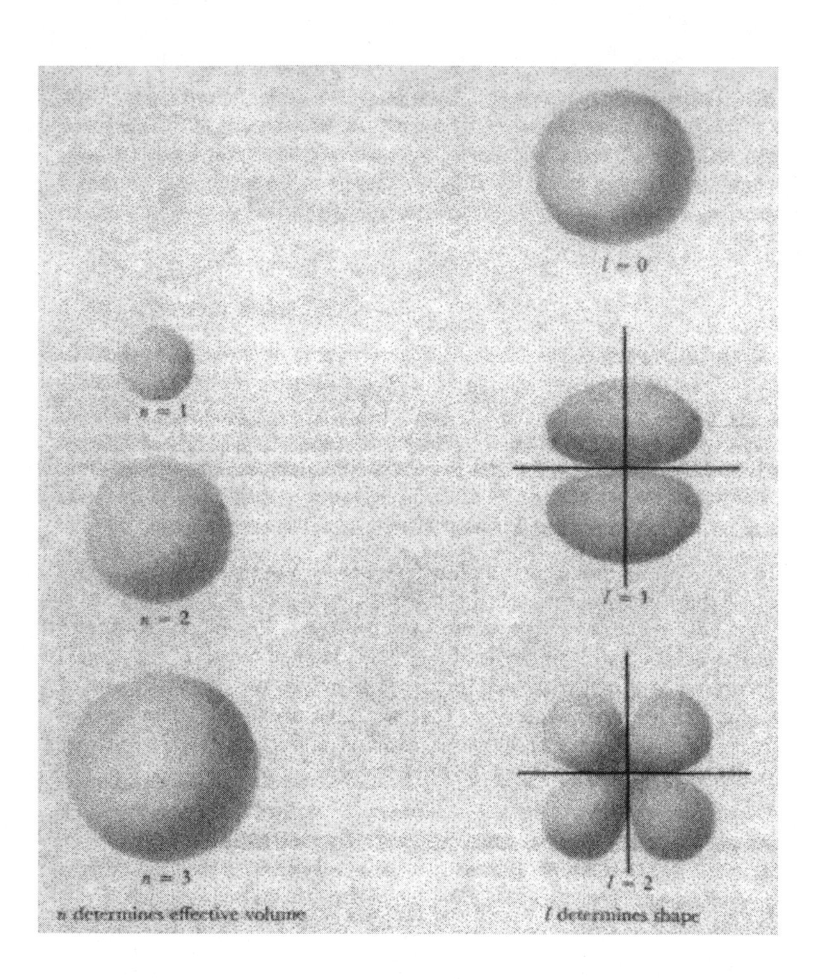

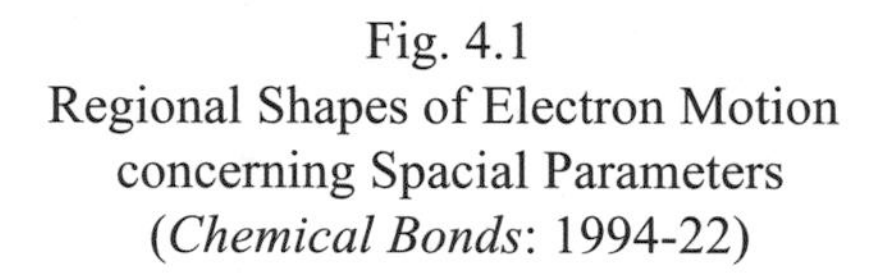

Fig. 4.1
Regional Shapes of Electron Motion
concerning Spacial Parameters
(*Chemical Bonds*: 1994-22)

Another is: one gram of hydrogen will combine with three grams of carbon to form methane. This was important because it was a consideration of the weight of an element as well as the atom of that element that was under exposition. Using this method, the weight for an element could be identified

Table of the Relative Weight of the Ultimate Particles of Gaseous and Other Bodies[36]

Hydrogen	**1**
Azôt [Nitrogen]	**4.2**
Oxygen	**5.5**
Water	**6.5**

Fig. 4.2

Relative Atomic Weights
(The Concept of an Atom-
From Democritus to Dalton:1991-95)

as that weight in combination with hydrogen. Richter's conclusion was still weak insofar as an explanation for elements that can form more that one weight ratio. Carbon is an example that will cause Richter's conclusion to be less than the only answer, because of the possibilities of more than one oxide being formed. Additional work along the road of multiple compounds and atomic combinations in this effort eventually came to the attention of Mr. John Dalton, who enhanced this thinking greatly.

John Dalton (1776–1844) thought a great deal on the particles that influence us and, in a paper called *On the Absorption of Gases by Water and Other Liquids* (1803), the focus of his thinking was primarily on particles and their behavior among themselves. He mentions this direction through the pressure one particle of gas would exert downward through a multiple horizontal strata of water until that particle of gas reaches a sphere of influence that another particle of gas would create. This area of thought is definitely in the discipline of pressure and density. It also is a reminder of Sir Isaac Newton's Proposition 23. This proposition was a great influence on Dalton's conceptions regarding atoms and their spacial qualities (fig. 4.1). In Proposition 23, Sir Isaac Newton considered that the distances between macro identities held a relationship to the distances that would be found in the micro dimension. Newton in his presentation considered an exactness in the distribution that precluded an exactness in spacial densities and would be useful in the determination of pressure relationships as well. Although Dalton rationalized that distribution of downward pressure is analogous to a pyramid concept and that all gases are not equal in pressure to water solubility relationships, his geometric model would still not lend itself to the immediate answer of particle weight. Additional thought to relative densities that could be expected was also on John Dalton's mind in that there had to be some definitive method through atom combinations wherein densities could be established as well.

In review of atomic weight combinations with respect to multiple proportions, John Dalton succeeded in extrapolating Richter's work and defined a method that made possible the relationship of an atom to its mass. From that conclusion, an atomic weight could be established. Figure 4.2 will establish the weights of four elements by Dalton officially called atoms by his friend Dr. Thomas Thomson in 1807.

The first known date on which a table of atomic weights was first presented was September 6, 1803 (McDonnell, 1991:89). John Dalton through his forward thinking on atomic structures and the relationships elements have on each other has given chemistry and the principles of atomic theory a very solid foundation. In essence, John Dalton's research and effort defined several areas that remain unaltered today. First, all matter is composed of atoms and all atoms of a like element are the same in the properties of weight, size, et cetera. Further, all atoms of different elements are different in their respective weights and size. Most important is his reasoning that no atoms may be created or destroyed in the chemical process; they are simply interchanged. In his statements of this rationale, he reflects that it is through the chemical process that atoms which are together may be separated and that unification from the bringing together of distant atoms is possible.

The Laws of a Gas

We should pause at this time with the knowledge that some effort had been given to the process and establishment of weight when concerned with an atomic structure. Richter's and Dalton's chemical expositions have clearly opened another door in the exploration of the atom. But how many atoms are in a specific volume? And when considerations of atmospheric constituents are involved, thinking along the lines of motion of these densities is very important, especially since there must be some correlation between how many oxygen and nitrogen atoms there are and the influence that they exert on us and our immediate surroundings.

First in any logical considerations regarding an atom when in association with atmospheric identities and historical reference is where would these two different atoms be and how many of them are there? We certainly feel something when the wind blows or we exhale. Observations relative to trees, paper, and whatever else will be subject to particle motion are clearly in evidence. Somewhere and in a time past these observations held interest for the scientific community.

Count Amadeo Avogadro (1776–1856) was certainly involved with these questions as they would apply to liquid and specifically a gas in a behavioral sense. In 1813 Avogadro, an Italian physicist, was intrigued by the studies done through the work of Guy-Lussac concerning the volume of gases in a chemical reaction.

In this particular investigation, there is a continual reminder that any atmospheric contemplation will involve, at some point, a gas. Gases will not behave like a liquid; they are considered to be special in an interactive way, and as we have seen with any atomic structures, they are governed by certain rules and physical laws. An ideal gas is one that essentially has zero mass because the particles are so small as to possess no volume and thus considered to be a point mass. In addition, the collisions are elastic. That is, there is no energy transference with respect to repulsion or attractive forces, so kinetic energy of a molecule is always constant. As far as pressure is concerned, when gas structures are in collision with a container's sides the force that is distributed by that particular side's area is by definition the internal containment pressure. That pressure is defined in atmospheres and will be explained in further detail later in this study.

Atomic structures in a gas will exit a certain property when the confinement volume is altered, and it was through the allowance of a gas to expand into another area of equally sized containment that Gay-Lussac noted a particular occurrence. As the gas expanded into what was an area that would effectively double the confinement area or volume, heat that could have been measured as a result of this process was absent. Interestingly enough, the first chamber's air temperature became cooler as the gas expanded into the second chamber. Also, as the gas began to expand into the second container's volume, the container's air temperature became warm. This observation generated a corollary between gas expansion and compression. Under compression, gas will increase the air temperature. Under expansion, the gas will cause the air temperature to become cooler, all without any addition or subtraction of heat energy from any outside influence or source (Williams, 1999:64).

The relationship of a gas pressure to temperature is made clear by this example, and in chemical circles it is known as Gay-Lussac's Law. Essentially, his law states that the pressure of a gas at a constant volume is directly proportional to the absolute temperature. Here the absolute temperature when gas pressure is zero and there is virtually no molecular kinetic dynamic is -273.15 degrees Celsius (-273.15C).

As far as how many particles of gas there are within a specific area, that is where Count Amadeo Avogadro makes the next step in the exposition of gas concepts a little clearer. He postulated, based upon the work done by Lussac, the explanation for the observations made with various gases was because in comparable volumes of the different gases under the same conditions of temperature and pressure there must have been the same amount of molecules present. This is Avogadro's Law.

The chemical term for a specific amount of grams equal to the nuclear weight of a substance of gas is a *mole* (Sill and Hose, 1968:468). In one mole of gas there will always exist the same amount of molecules; that number is 6.02 X 10 to the 23rd power. The Ideal Gas Law will reflect this in its equation property as having within it a constant (*c*) that will be representative with any gas. The equation PV = acT will then work for any desire in solving for the fourth property if the other three are known (1966:122). It is understood that the constant *c* will be the same for all gases and given when moles are used as R or the gas constant. Therefore, the Ideal Gas Law may be expressed as PV = nRT. Here *R* is not then dependent on the type of gas studied, because equal volumes of gases under the same temperature and pressure will contain the same amount of molecules.

The Molecule

An atom is the smallest known particle of matter. The molecule is comprised of atomic heavy nuclei or ions, and the composition of "kinds" of atoms or nuclei may be different (1969:246). Molecules may exist in a great many ways; one in particular of great interest is water. Another unique property of the molecule is because it is comprised of atoms that are bonded together, there must be some space between them within the substance. Under pressure, the distance between the particles will lessen and become smaller. When these conditions are involved, a liquid may be changed into a solid. These are just a few of the many interesting prop that a molecule will have. This particular aspect of the chapter will d with these and a few more facts in the construction of a knowledgeable foundation, regarding the mechanical structure, binding forces, and electrical quantities, that will eventually be recognized later as the atmosphere.

First, not all molecules are similar in their respective molecular geometric bond configurations. This observation can be made between two of the simpler substances of carbon dioxide and water (fig. 4.3). The atoms of water vapor will have a triangular configuration with the oxygen atom at the apex. Because of this type of structure, orthogonal rotations about the axis will be three in number and pass through the center of molecular mass. Moments of inertia with this consideration will all be different accordingly. Geometrically, there are several types of molecules in specific categories in the way they are structured. Figure 4.4 illustrates these typical geometries. Not all molecules will have the same stabilities and lifetimes, some molecules will last for less than a nanosecond while other molecules will have a lifetime that is measured in the billions of years.

The classification of a molecule may depend on the amount of atoms that are required to bond together in the formation of the substance. In cases where only one atom is required, that classification is monatomic, two would be classified as diatomic, and three classified as triatomic. Molecules with a greater containment of atoms beyond three are classified as polyatomic. It is a fact that in larger molecules such as diatomic structures, every kind of atom will be found in at least one of them (1990:761). As a general rule, molecules may be identified by their specific atomic containment, and these would include the electrostatic charge, if any, that some may possess. The way in which the atoms are joined together or the binding configuration is also important when discussions concerning electromagnetic properties are involved. This portion of the molecule is very important since the electrons in orbit must be shared between nuclear masses for a structure to be stable.

With regard to the bond type and configuration, it is recognized that because a bond of significant strength exists, there is a force represented that defines that holding property and will be covered later. Some cases in a molecular configuration may require several bonds between atoms; these situations are regarded as double or triple bonds. From a schematic viewpoint, the bonding type will be drawn as either one or a multiple number of lines drawn between the respective atoms. Although the lines will give an indication of possible strengths, they will not, however, indicate spacial,

Tri-Atomic Vibration Dynamics

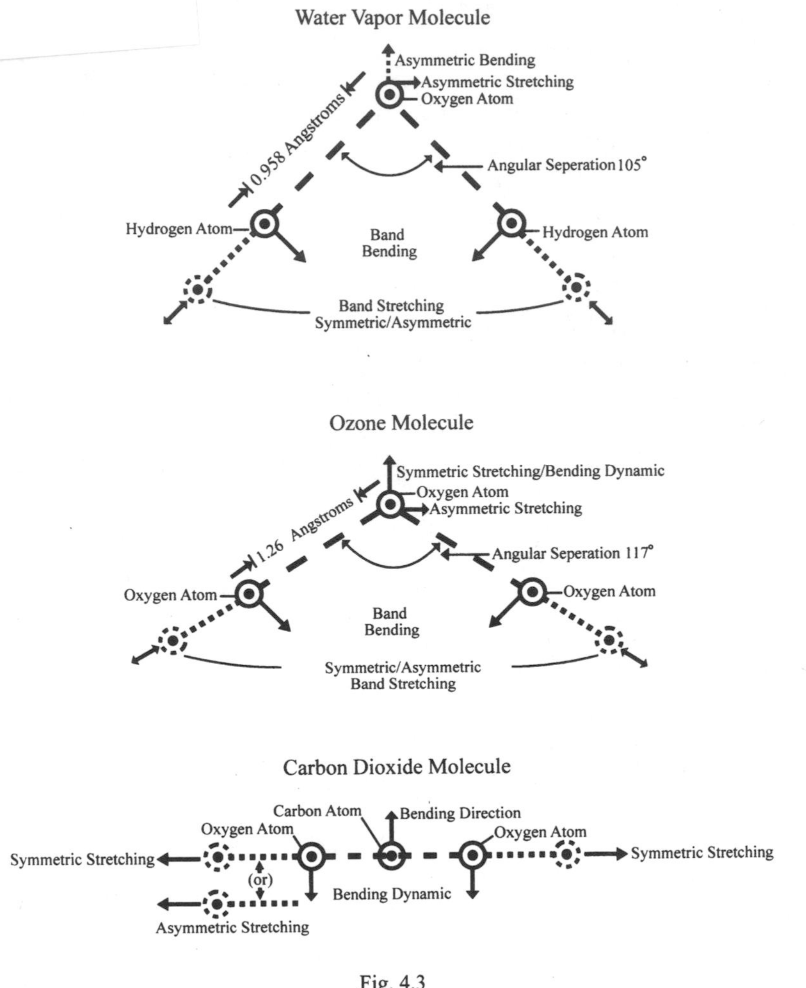

Fig. 4.3
Inter Bonding Structures
Interpretation of Text
(Mendes-Hussey Graphics)

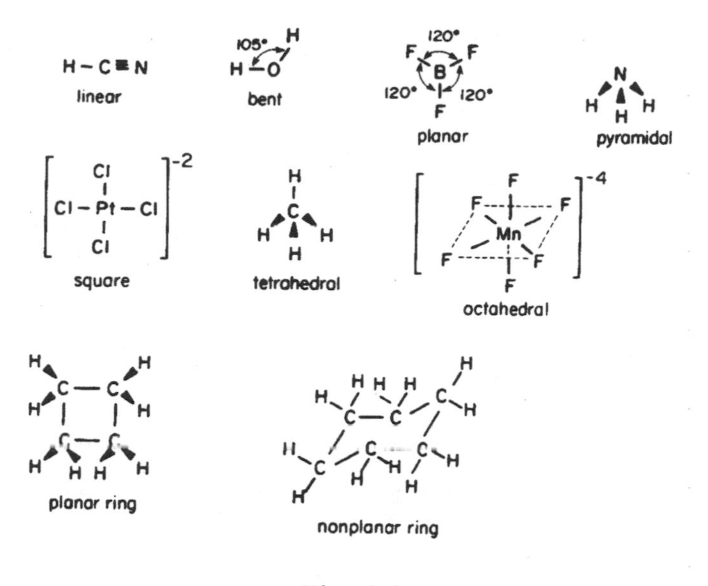

Fig. 4.4

Typical Molecular Geometries
(Encyclopedia of Physics:1990-762)

length, or angular orientation of the structure. In a structural isomer, for example, although the same elemental compositions are identical, the bond structures will differ.

In the molecular bond process, the individual atomic electrons play a significant part. With this in mind, a bond will form when the distribution of the combined electron totals is of a lower energy than if the electron energies were totaled separately with their original atomic structures. One example of this union is the covalent bond in which the original two atoms each have one electron in orbit about the nucleus. When combined in a molecular structure, both electrons are shared within the covering cloud. Although the repulsion distribution is more enhanced, the kinetic energy of the total combination will be lower (1990:763). Covalent bonding is generally recognized in configurations whereby the previous electrons were only in a single orbit. In cases where multiple electrons were in a single orbit about their respective nucleus, then the possibility would exist in the formation of a double bond, and in the case of three electrons within a sin-

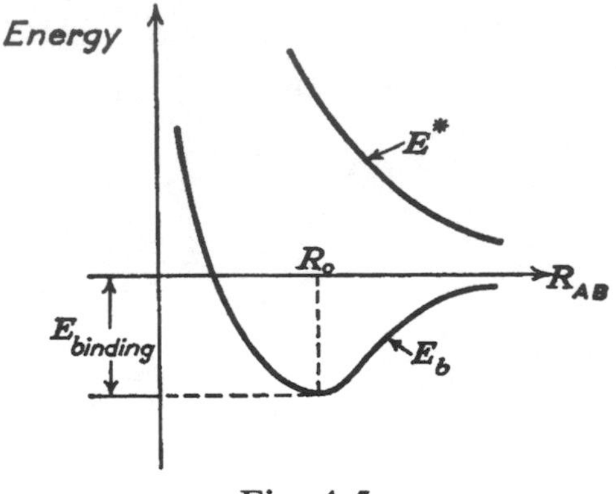

Fig. 4.5

State Energy as a function of Protonic Distance
(Atomic Physics:1969-273)

gle orbit the formation of a triple bond would be possible. If an electron transfers into an orbit of lower energy from its original previous orbital higher energy state to an atom that has an electron missing, that transfer of the electron into the unoccupied orbital position would constitute an ionic bond. The valence shell or outer atomic orbits will dictate the covalent bond possibilities, for it is when the outer shells are filled with the full complement of electrons that ionic bonding will take place. It is this form of electron delocalizing that makes the ionic bond more susceptible to energy lowering or termed resonance energy and therefore provides a more stable molecule.

The types of molecular binding will vary because of electron valence occupation. When the valence of the atom is saturated, that is, all valence electrons are present and full occupancy is achieved, that is termed *covalence bonding.* Most of this type of binding situations belong to the gases of H_2, N_2, and O_2 and most organic compounds. With respect to covalence bonding, the H_2 molecule can be discussed to provide a rather secure understanding of the process through which the binding and capture may take place.

As an example, the H_2 ion or atom with a single electron in orbit about the nucleus can be presented. When the two protons are rather far apart or, in this case, a few Bohr radii, atomic wave functions will be acting individually on each proton accordingly. At this time the electron will be affected by each proton equally or in the sense that it will spend time in orbit about both, and this leads us to a defined term of a molecular orbital. If, however, the proton's distances (RAB) become closer, the energy (ENERGY) of the system will also decrease. The equality of charge potential is redistributed into a region between the two protons, and they become somewhat screened from each other. As the proton's distance closes further, eventually the screening effect (equilibrium [Ro]) is broken and the like repulsion force between them will effectively drive them away again. Here we can see that as long as the screening effect is in place and the energy is at its lowest state, the bond will hold (1969:273) (Fig. 4.5).

Those bindings that will form a lattice structure will be termed *metallic bindings* and can usually be found in most metals. At the molecular level, when a dissociation occurs to break apart the atoms of a molecule into their respective neutral states, as in thermal excitation or as a consequence of a radiative process, that is termed *atomic binding* (1969:267). Obviously, the rate or ease of extracting an outer electron varies; however, the farther out in an orbital configuration an electron will be, the easier for that electron to be influenced into a transference situation or capture. That particular ease to which the electron is subjected to be removed is the ionization potential.

Interatomic Forces

It is understood that when a difference of electrical potentials exists that creates a deformation of the electron cloud in such manner as to create a bipolarity to what was previously a nonpolar arrangement and when that electrically polarized molecule influences others of similar polarizations, then the potential of Van der Waals forces is in effect (Seinfeld and Pandis, 1998:667).

These cases are apparent in molecules that will be attractive to each other without being chemically bonded at large distances and exhibit a repulsive action at short distances. This force is generally used to describe an influence that is different from adhesions predominate in ion or valence molecules (1990:544). Importantly, it is this specific force that in most cases will explain the existence of matter in the states of liquid, solid, and gaseous forms that can be found with high degrees of stability in a variety of temper-

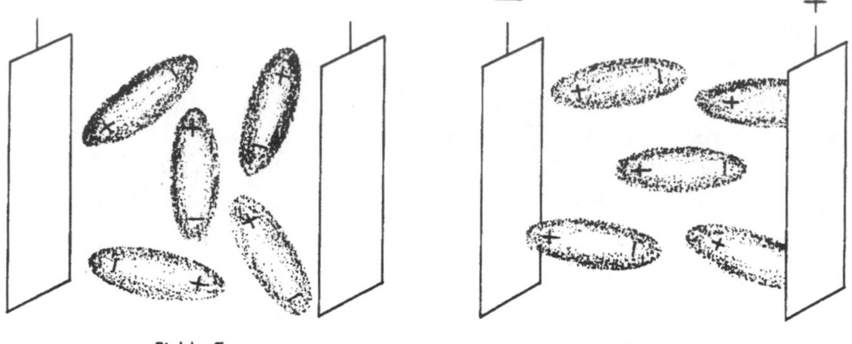

Fig. 4.6
Inductive Alignment Influence

(Chemical Principles:1966-217)

ature and pressure existences. Repulsive forces are an integral part in the formation and final configuration of the molecule itself, because of the fact that they are more profound at distances that would be equal to or less than the smaller molecule's diameter and define the eventual molecular configuration. The principle reason for this activity is the electron to nucleus relationship. Range influences are much smaller in a nonelectrostatic condition than would be found in an atomic relationship that was predominately electrostatic. Although the polarity or dipolar configurations are of importance when discussions of molecular arrangements and Van der Waals forces may be concerned in crystals, they themselves do not play a significant role in the eventual binding energies of solids (1990:544).

As the electrons exert an electrical influence about the nucleus because of their dynamic motion, that motion will in effect establish electrical fields that will be an inductive influence for another molecule in proximity to it. Because a field is created by the electron dynamic, it is that influence on other clouds that will create an alignment of that neighboring molecule in such a way as to create a dipolar orientation called the dipole moment (fig. 4.6) (1966:217). This inductive process through the electric dynamic or Van der Waals force is therefore a significant portion of the binding influence within certain molecules.

As previously learned, larger molecules with electron valence arrangements contained in outer distance shells will be subject to the induction much more readily than that of molecular arrangements where the electrons are much closer to the nucleus and would not necessarily be disturbed by the proximital transient of an electric field. Also, because it is the electron's shift that would produce a bipolar situation, the larger the atomic weight of the atom, the larger the Van der Waals force would be. In the case of the water molecule, because of the hydrogen atom having its two electrons tight agent the nucleus, the Van der Waals forces are only representative of about twenty percent of the bonding action. Hydrogen bonding is eighty percent (1966:237).

Although the predominant force of influence between nonpolar molecules is Van der Waals, two other forces are operative, one of which is stronger as an attractive influence in a bond relationship, the hydrogen bond. This bond is the strongest of most dipolar forces because of the fact that the hydrogen molecule will act almost as a proton in its relationship in a binding force influence. Hydrogen's small size will allow close proximity to its nucleus by the approachment of several other molecules. It is this close nuclear proximity that results in the almost protonic electromagnetic effect, thus creating interatomic bonding attraction and retention. The most influenced molecules are those with small nuclear diameters and include fluorine, nitrogen, and oxygen.

Results show that the larger molecular structures do not have the effect that these other mentioned molecules exhibit insofar as hydrogen's nuclear attraction goes. This is because of their large radii, and so they tend to not be as influenced into bonding. Water is one of the results of a hydrogen bond and is explained through the high surface tension and relative viscosity manifested. In addition, because of the almost protonic effect hydrogen has on smaller nuclear radii, the heat capacity of water is an example of the high heat requirement necessary to break the hydrogen bond.

Molecular Bond Dynamics

As the atom is combined into a molecule, its structure will have certain geometries that will have motion with respect to each other and to the nucleus through the bond relationship. The molecule may manifest motion (rotation) as well as the individual atoms (vibration). This study into bond

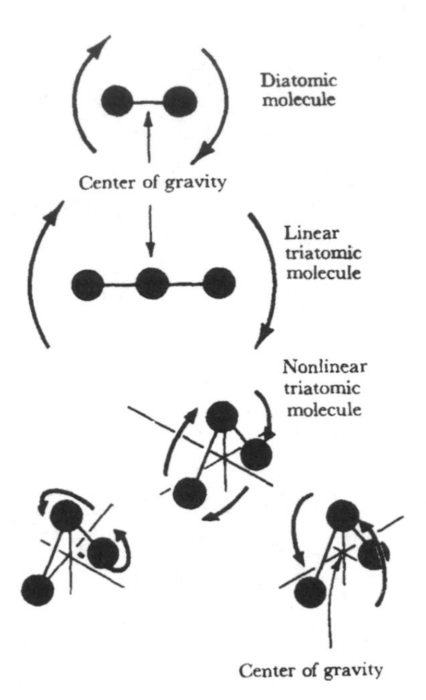

Fig.4.7

Rotational Molecular Motion Dynamic
(Chemical Bonds:Am Intro. to Atomic
and Molecular Structure1994-136))

dynamics is important when concepts such as radiation are used as the impetus for such movement (fig. 4.7).

Let us call the dynamic of the molecules impetus excitation. As with most molecular geometrics, position of its electrical possibilities would always be of concern. Since a molecular structure can be deformed electrically, that is to say it is polarizable, it may be in any of innumerable positions prior to contact with an electric field. Also, when the structure comes into proximity of an electric field a realignment is to be expected. It is this susceptibility to a spacial alignment from an electromagnetic stimuli that is called a molecule's state of excitation. The resultant nuclear dynamic that is created within the geometric confines of the structure itself is

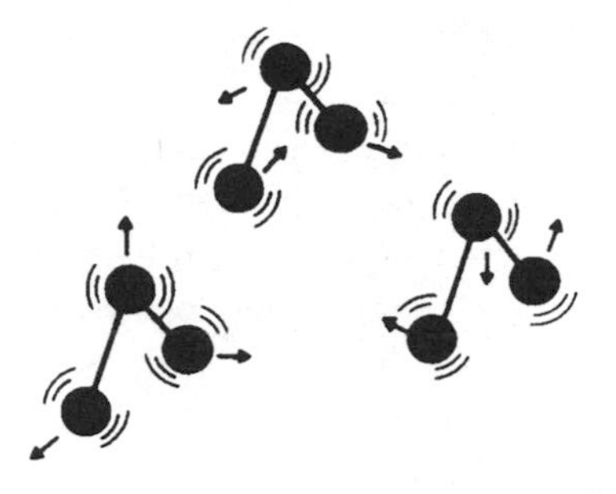

Fig.4.8
Vibrational Motion Dynamic
(Chemical Bonds: An Intro. to Atomic and Molecular Structure:1994-136)

identified as its motion. That particular motion is either rotational, as in the case of molecular geometry or vibrational, as would be in the case of its individual atomic elements.

Because the fact that other elements are attached at some distance from each other gives way to allowable movement, that can be observed as being independent between nucleons. In the case of a diatomic molecule, because a distance exists between the nuclei, a rotational motion can be developed that would cause particles to be about a center point or common axis. The dipole may have dynamic motion that causes the two nuclei to move in opposite or closer directions, or the motion in this case would be considered to be vibrational (refer to fig. 4.3). Not all diatomic motion is the same across the atomic spectrum; if emission from the atomic geometry is measured, then results show behavior that is relative to individual characteristics identified by specific vibratory states. Additional vibration states can also be specified by angular distinctions pertinent to individual molecular geometries, such as the cases in figure 4.3 identifying structures in the triangular formations. Electromagnetic characteristics of molecules as far as dipolar susceptibilities are concerned are the principal reason the geometry will be excited into dynamic motion, only if the result of that motion is an oscillatory dipole moment. In cases whereby a polyatomic molecule is excited, the rotational aspects of that geometry will vary with the different moments of inertia about the varied axis possibilities and is characterized in a different way (1969:258).

Rotations from the standpoint of simplicity can best be applied to a molecular geometry that has a single bond between particles. In this geometry, flexing, torso rotation between particles in that one particle will twist and another will not gives a great deal of freedom to dynamic possibilities.

An illustration of those possibilities that show the myriad of rotational motion can be seen in figure 4.4, Please notice the differences between molecular rotational and vibratory states. When a rotational situation occurs, the spin and motion are about the central point of a linear polyatomic structure along any of three perpendicular axes that are mutually part of the whole geometric molecule (1994:135). The entire molecule will rotate, whereas in a vibrational state each atom will shimmy and move toward or away from another attached atom; however, the entire molecular behavior is confined to the end points in relation to each other's attachment point (fig. 4.8). These possibilities will be in effect, the radiative and hydrological properties that atomic dynamics make possible within our atmospheric envelope and which will be covered in some detail later.

5

Construction of an Environment

Atomic Implications

So far, we have studied the Law of Gases and those individuals responsible for the discoveries that have led science to understand that we are surrounded by atoms in the form of a gas. We have also studied the atom to its smallest detail and the motions that its electron cloud will undergo when subject to electromagnetic fields. The next foundation to build upon is an important one and takes the form of a question that may be expressed in several parts. That question is: What atoms and in what proportion is their existence in the form of the context of a gas? In addition, does the gas become visible and what are its behaviors that would be of such importance to us as a species we would care enough about to document. Because after all, isn't that what would be relevant and purposeful behind an effort to chart something?

Rutherford's credit with the atom of nitrogen within our gaseous surroundings, Priestley with the discovery of the carbon dioxide atom, and Antoine Lavosier with the discovery of the oxygen atom all had something in common. All the atoms of discovery existed within an environment that is seemingly invisible to us, called by Johannes Baptista van Helmont in 1748 gas. Today the encompassment of those atoms is termed *atmosphere*. How much of the totality each comprises and where the individual atoms themselves are located can be determined by the exposition of their densities at variable heights above the ground. The variations of atom densities can be attributed to the nomenclature of turbulent air, which has the mechanism of mixing the various constituates under the altitude of 100 kilometers most equally. The region of mixed atoms below 100 kilometers is called the homosphere (1996:11). Figure 5.1 illustrates the densities of constitute atoms as a function of altitude.

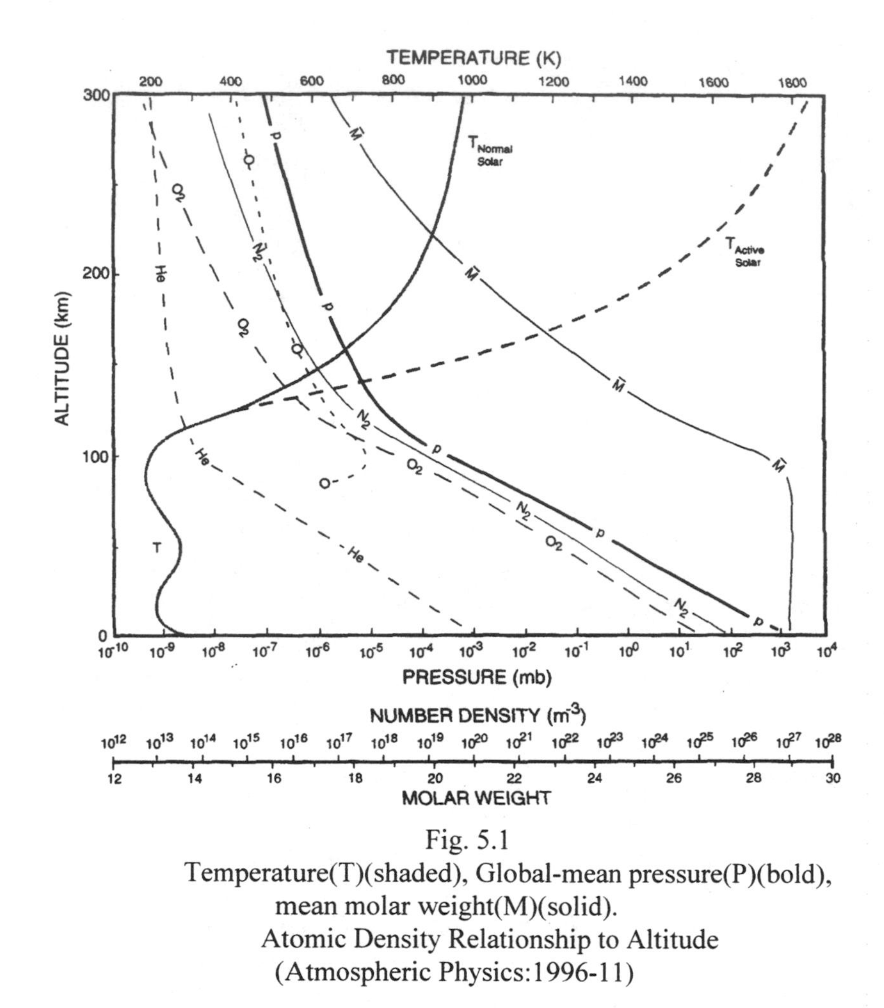

Fig. 5.1

Temperature(T)(shaded), Global-mean pressure(P)(bold), mean molar weight(M)(solid).

Atomic Density Relationship to Altitude

(Atmospheric Physics:1996-11)

As for the individual analysis of each atomic structure, the heaviest is O_2 and that will decrease more readily with altitude, N_2 is next and so forth within the heterosphere or the region above 100 kilometers but less than 500 kilometers. Any investigation of the atoms and how they form densities within the homosphere should be done with a context of origin. That origin, insofar as what was around and when, is vital to the precedence of what is in existence now. That is, what was then may not be what is now.

The early envelope of most likely atomic particles in abundance such a long time ago first existed as hydrogen and helium. As the Earth cooled and the interior sections of magma formed the rock litheosphere of our planet, volcanoes outgassed through the eruption process, venting water vapor, carbon dioxide, and a very small percentage of nitrogen. The reason for the high percentage of nitrogen within our surface atmosphere today is thought to be carbon dioxide concentrations' reduction by absorption and dissolution processes, biologically and chemically becoming internally held within formations of terrestrial limestone (Ahrens, 1994:7).

Although oxygen represents about 21 percent of the surface homosphere, its origin was considered to be a result of solar influences breaking apart or the process of photodissociation concerning the hydrogen and oxygen atoms, existent as the water molecule at that time that formed the new oceans and bodies of water.

The atmosphere or, in meteorological terms, troposphere is closest to the surface of our planet. The meaning is "turning sphere," which identifies it as the most convectively active, which is precisely why the atomic constituents become mixed rather equally. Its highest point is 100 kilometers above the surface bordering a thin layer called the tropopause, which is the buffer between the troposphere and the next layer, called the stratosphere (1996:16).

The current model of the nitrogen molecule has ten valence electrons. The bond arrangement is triple, with an N–N atom distance of 1.47 angstroms. The energy required to dissociate the molecule into its respective two atoms is very high, which is a result of that triple bond. Also, because of the bond strength, nitrogen atom reactions are minimal with respect to reactions of other elements (1966:246). As with any consideration of atomic or molecular dissociation, in a stronger bonding arrangement, since an overall energy balance is between the absorption of the old bonding and the new as bonds that are broken, an instability will occur that will influences any new bonds. Because of the high-bond energy of nitrogen, any new bonding will have a very little balance to work with, and as a consequence, the low energy difference makes for poor joining and thermal stability prospects with other elements or compounds. Through chemical analysis, either the nitrogen atom or molecule will react with only two other nonmetal elements in a very direct way; those two elements are hydrogen and oxygen, which appear most abundantly as bonded molecules in the same manor as the nitrogen molecule (Danielson, Levin, and Abrams, 1998:39) (fig. 5.2).

Fig. 5.2
Typical Atmospheric Constituates
(*Meteorology*: 1998-38)

Oxygen

The oxygen atom and the diatomic molecule it forms within the tropospheric envelope began about 1.8 billion years ago as a by-product from the new formations of plant life producing it through an interaction with the sun. Today we know and define that process as photosynthesis. Oxygen by atmospheric volume is 21 percent and is still created by that process. Being heavier in mass along with the outgassed molecules oxygen was retained by the Earth's gravity and so settled closest to the surface. Scientific evidence in geologic formations will confirm that oxygen was formed such a long time ago and is apparent through iron depositions in the terrestrial surface rock.

In the normal state within the atmosphere, oxygen atoms are generally found in the bonded configuration of a diatomic molecule and referred to as O_2 (1966:243). It has twelve valence electrons in the orbital shells around the nucleus, with the two atoms held together by a double bond. Because the oxygen molecule has a tendency to be attracted by an electromagnetic field, that is, it is paramagnetic, it is generally understood that the

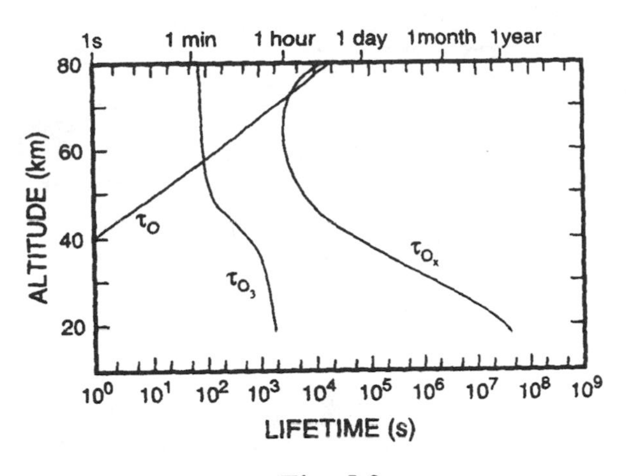

Fig. 5.3

Oxygen Species Photochemical Lifetimes
(Atmospheric Physics:1996-537)

molecule contains two unpaired electrons. It must be noted also that oxygen has a weak Van Der Waals force and will be a gas at room temperature and standard atmospheric pressure.

Additional existence of a different form of oxygen molecule known as ozone will occur if an O_2 molecule is subjected to an electrical charge. The new formation of the O_3 molecule or ozone is then created. When a situation arises whereby an element can exist in two forms or molecules, that is known as allotropy. In this case, both ozone and oxygen are considered to be allotropic forms of that element (1966:244). This particular molecule will be scrutinized much closer when the subjects of radiation and subsequent reactions are discussed later in this book.

As with all elemental existence in combination with something that initiates a reaction to the atomic configuration, the word *lifetime* is an important one. The O_2 molecule can and does undergo a dissociation process when bombarded by ultraviolet (UV) radiation, termed *photodissociation* or *photolysis*. The O_3 molecule will be similarly broken into atomic oxygen (O) and the molecule (O_2). At this stage, the atomic oxygen then recombines again with the O_2 molecule in another ozone configuration. The lifetimes of the individual atoms and molecules may be referenced in figure 5.3, and attention should be drawn to the particular function of altitude

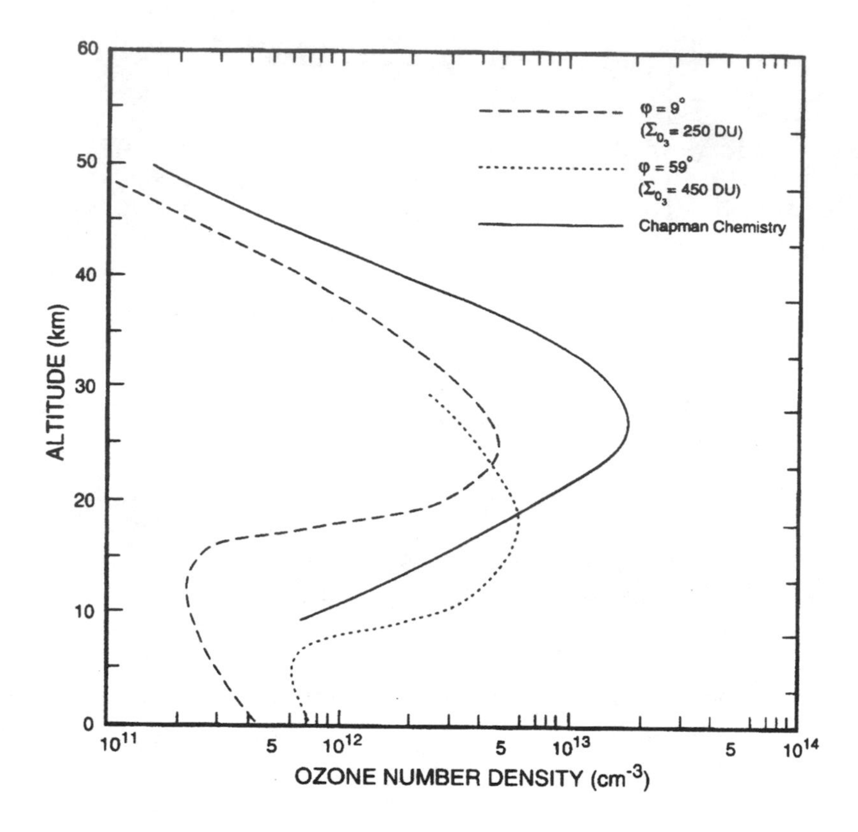

Fig. 5.4

Ozone Densities as a function of Altitude
(Atmospheric Physics:1998-540)

(1998:536). Because the oxygen molecule is susceptible to UV radiation, it is normally expected that with the absorption of that wavelength, the reaction would be the dissociation of the O_3 molecule. At lower altitudes, because the higher regions have already removed some of the UV radiation, an increased lifetime would be the expected result for O_x molecules. The fact that the O and O_2 configurations are in existence for such a different period of time would suggest that a photoequilibrium is in effect. This suggestion is verified through the illustration of figure 5.4, and will also show that the evidence is strongly presented to indicate the tropospheric convective nature of transporting the O_3 ozone molecule through a variety of altitudes below thirty kilometers where photosynthesis processes are taking place.

From a climatological viewpoint, the most favorable situations for the formation of ozone would be as a result of a warm slow-moving high pressure system whereby through the cloudless atmosphere the sun's radiative effects of mixing would be most efficient in the photochemical process. In a case where a cooler, lighter dynamic and a cleaner air is found over city regions, these particular combinations increase the detrimental effect to ozone production through an atmospheric inversion consequence. In these situations, the temperature at lower tropospheric altitudes will increase with height and will not allow the cooler air to mix with it. Thus the amalgamation of cooler air does not occur (Seinfeld and Pandis, 1998:262).

As far as the amount of ozone is concerned, 90 percent of it is found in the stratosphere and about 10 percent in the troposphere (1992:436). The totality of ozone, detrimentally speaking, has decreased 3.5 percent, determined through analysis of satellite imagery from the Total Ozone Mapping Spectrometer (TOMS). These data were extracted between 1979 and 1989 between the south latitude of sixty-nine degrees and the north latitude of sixty-nine degrees. Previously, ozone losses were the greatest in the spring and winter seasons; however, instrumentation will now confirm that stratospheric losses are occurring throughout the year (1998:93). Although ozone is transported from the stratosphere into the troposphere by what is termed *tropopause folding events,* which are defined as tonguelike intrusions from the higher stratosphere into the troposphere below it, the amount of intrusion is only about 1 percent of all stratospheric production. In view of stratospheric production losses and the small intrusion factor, the balance of ozone from higher levels as any replacement consideration to lower tropospheric photosynthesis creation is insignificant. Just as the stratospheric ozone levels are diminishing, tropospheric levels have increased over Europe by between 1 and 3 percent over the past thirty years (Seinfeld and Pandis, 1998:94). In addition, the data from TOMS suggest that from 1979 to 1992 the amounts of Pacific South American amounts increased by 1.48 percent per year. This value was significant to a total increase over the Northern Hemisphere as a result of NO species emissions (Seinfeld and Pandis, 1998:94). Because emissions in the Southern Hemisphere concerned were related to biomass releases from burning, ozone (O_3) is more predominant.

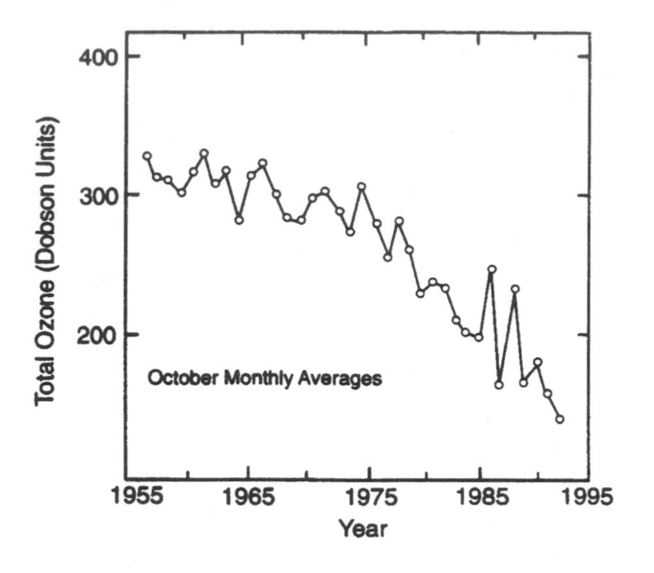

Fig. 5.5
Historical Trending at Halley Bay, Antarctica
(Intro. Atmospheric Chemistry:1999-180)

The destruction of ozone from anthroprogenic emissions can be identified as coming from chlorofluorocarbons, or CFCs. With large-scale upward convection at the tropics as the transport mechanism, when CFCs reach stratospheric levels a decomposition into free chlorine atoms takes place. It is these free chlorine atoms that will react with the ozone molecule in its destruction. It is most worthwhile noting here, since CFCs have a lifespan of about one hundred years (1992:439), ozone depletion from this molecule will not dissipate any time soon even if anthroprogenic emissions are curtailed. As far as a significant expended state is concerned, since Arctic regions are inherently cooler than other temporal locals, the photochemical depletive process had been anticipated to be slowed; however, there is excellent speculation about greenhouse gases in the stratosphere having a cooling effect, as opposed to the heating effect observed in the troposphere. This cooling effect within the stratosphere can be attributed to the radiating of any heat that would be associated with O_3 molecular absorption of UV radiation (1999:187). Hence, sadly, the trend is still a negative slope (fig. 5.5).

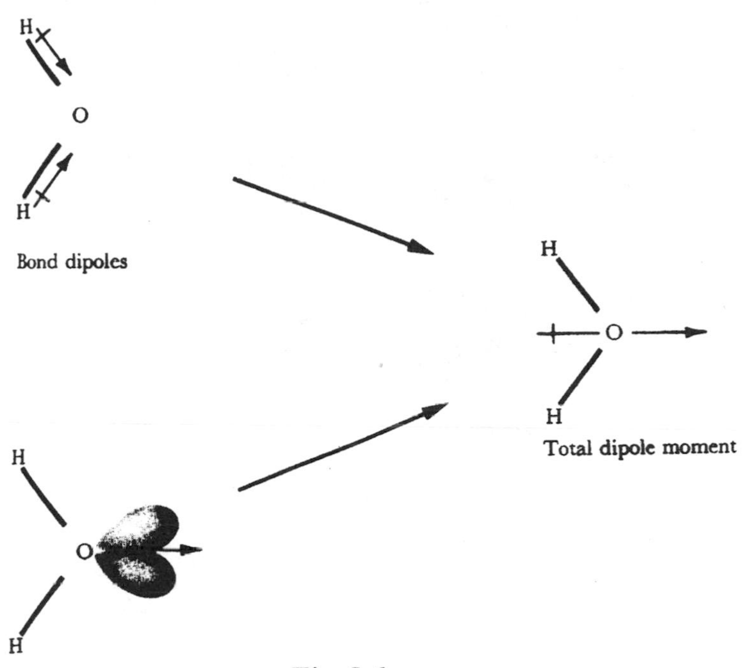

Fig.5.6

Electronegitivity Dynamic
(Chemical Bonds:An Intro. to Atomic -
and Molecular Structures:1994-125)

Water Molecule

Of all the molecules that the atmosphere contains, this molecule is probably the most fascinating to study as an atmospheric constituent, because of its predominance in such a large part of our lives. First, the H_2O molecule is polyatomic and polar; that is, the valence orbits of the oxygen atom of are a lower energy level than that of the hydrogen, and because of this electrical imbalance between the hydrogen and oxygen atoms the electron pairs of the two oxygen atoms are a major negative influence in creating a motion that results in an attraction with the electrons involved in the O–H bond to move toward the oxygen atom as illustrated in figure 5.6.

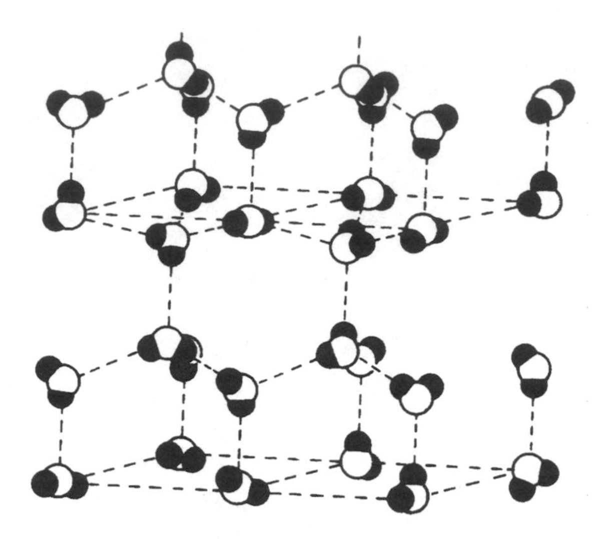

Fig. 5.7

Water Molecule as a Solid Structure
(Chemical Bonds: An Intro. to Atomic-
and Molecule Structure:1994-204)

Also, precisely because the oxygen atom will consist of two lone pairs in the hydrogen-to-oxygen bonds, this imbalance will cause an attractiveness to an electromagnetic field with the consequence of the water molecule having a distinct polarity and dipole moment (Gray, 1994:124).

It is the electroquantums that make the hydrogen bond so strong compared to others. In the case of a water molecule, as previously mentioned, the bonds are very cohesive in both the solid and water phase of this molecule and, because of this type of bond with oxygen, the temperatures of melting and boiling are very high, especially when compared to other compounds that contain a hydrogen bond, such as H_2S, H_2SE, or H_2TE. It is interesting to know that because of an open network structure that is consistent with a hydrogen bond (fig. 5.7), which is tetrahedral, when melting does occur, not all the hydrogen but about 28 percent of the bonds are broken. As a conclusion, one can state that some of the ice structure will always be present when water is in its first stage of liquid form. As the temperature rises and the tetrahedral structures collapse, the greatest density that liquid water may contain is at four degrees Celsius.

Water may be observed in three states or phases, the first being a solid, which we call ice, as described earlier, in which case the molecules are arranged in a specific order and although they may still vibrate, the individual molecule is secured and unable to move freely. The second phase that the water molecule may be in is as a liquid. Unlike in the solid and fixed position, the molecules move about somewhat more freely and, being very close together, will bump and bounce against one another. In this case, at the very surface of the liquid, if seen in a magnified way, the observation would be made that not all molecules are moving at the same speed. Some may be in motion at a sufficient velocity to bounce off another and move into the atmospheric space above the liquid body. This breakaway situation is termed *evaporation,* which may be described as a cooling process and related very closely to the kinetic energy of the molecule, which as a liquid will have a known average kinetic energy at a given temperature (1966:265).

Just in the same manner as molecules of water are leaving the surface area, some molecules are arriving from the atmosphere above and around the liquid body. The attraction of water molecules in a liquid state is, as expected, greater, in the interior because of the density of surrounding molecules, than a molecule on the surface would be subjected to. Because of the greater attractive pulling effect on the surface molecules, a surface molecule of water will have an influence to be attracted to the center of the density. This is identified in a law of physics that states that an intermolecular force of attraction between two molecules is inversely proportional to the distance between them. Since water in liquid form has molecules close together, the attractive force would be very strong, as opposed to molecules having a greater distance between them. It is for this reason that liquid will try to possess the smallest surface possible and, when falling through a liquid that will not mix, such as water, the smallest possible surface would be in the form of a sphere (1966:265). When a gas molecule is returned to the liquid state, that is condensation and will occur as the molecules are cooled, causing a slowing effect in their motion dynamic, which results in attachment and adherence to another. When a balance is achieved in a deposition and evaporation process resulting in neither the liquid nor atmospheric quantities being enhanced, that balance is known as saturation (Ahrens, 1982:113).

In the gas phase, the molecule is completely free in its motion and will move about the atmosphere or containment from the kinetic energy of it, bounce, and come into contact with other molecules of nitrogen, hydrogen, and oxygen that exist also in the free atmosphere around us.

In a situation where the water molecules pass directly from solid to a gas, that process is called sublimination. In the circumstances of a water molecule coming into contact with a solid ice arrangement and being affixed to the ice structure, that process is termed *deposition* (Ahrens, 1982:112). All these activities have a relationship to one another insofar as a total system may be arranged. That system will at times consist of dispersed water molecules in all three phases at one time, and commingling with the water molecule are oxygen atoms, oxygen molecules, and a greater percentage of nitrogen atoms along with their diatomic nitrogen molecules. All the preceding within the troposphere and stratosphere comprise a molecular envelope that is called our atmosphere. How they react with sunlight and develop wind and pressures consistent with climatological patterns is indicative of a composition exclusive of the atomic implications by their membership.

Atmospheric Relatedness

There are differences between an implication and the application of that implication's component. In this case, we have an implied reality by the virtue of atomic parts. The next step from an applied viewpoint is to take those atomic particles in their form of existence and, through observable facts determine if they constitute a structure that has a relative importance to the term *atmosphere.* By this measure, we can then use those findings to develop a methodology that will expose behaviors that would be measurable in sufficient quantities to chart.

To start, in order to establish a behavioral foundation, compare a known existence with a truth in physical law that is demonstrable. The rationality is to determine fact. Here the existence of atomic structures must be compared to laws of different reference positions. One reference position is that of temperature. We know that temperature exists. We know that variances in the value that we assign temperature, that is, in degrees of Fahrenheit, Celsius, and Centigrade, indeed vary at given times and that when such events do occur they effect the molecular dynamic. It is that change in the molecules' dynamic that will identify the amount of molecules in a specific area. Since gravity influences all the molecules in ques-

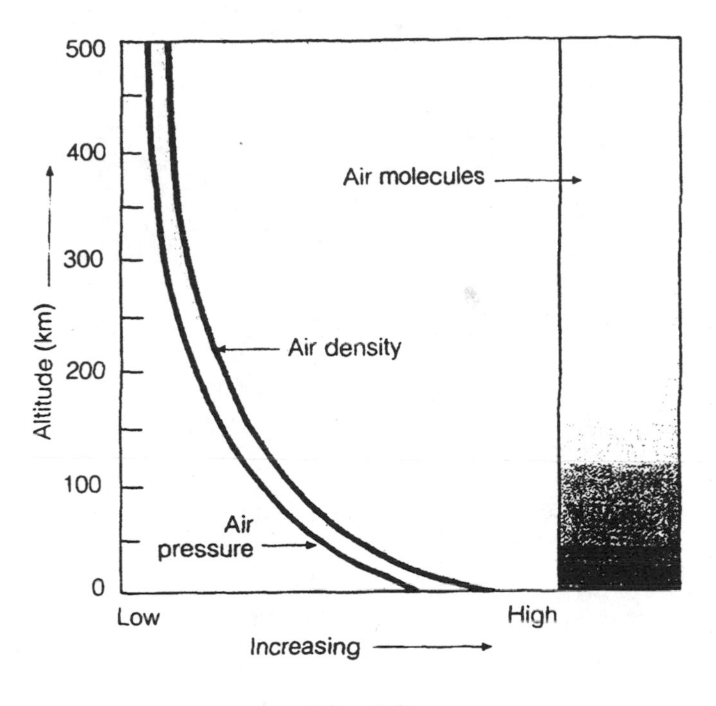

Fig. 5.8

Molecular Pressure and Density with Altitude
(Meteorology Today:1982-8)

tion and would affect the distances between them, than an identification of their amounts and location should give us an explanation of density. Since molecules have motion, they will exert a tiny force on everything they surround. That force against all things is called pressure. The truth of this can be found by observation from altitudes in that there are fewer molecules of the types mentioned at higher in altitudes or farther away from the surface at sea level as one takes measurement. The conclusion is that air does not exist in space, and this proves that gravity at certain distances away in height is a decreasing influence. The consequence of these facts will determine that with less gravity at an altitude, fewer molecules will be found and the greater the distance between them will be; that will be the truth of density and confirm pressure. Figure 5.8 illustrates how the truth of pressure and density is related in a law that governs an expectation from them.

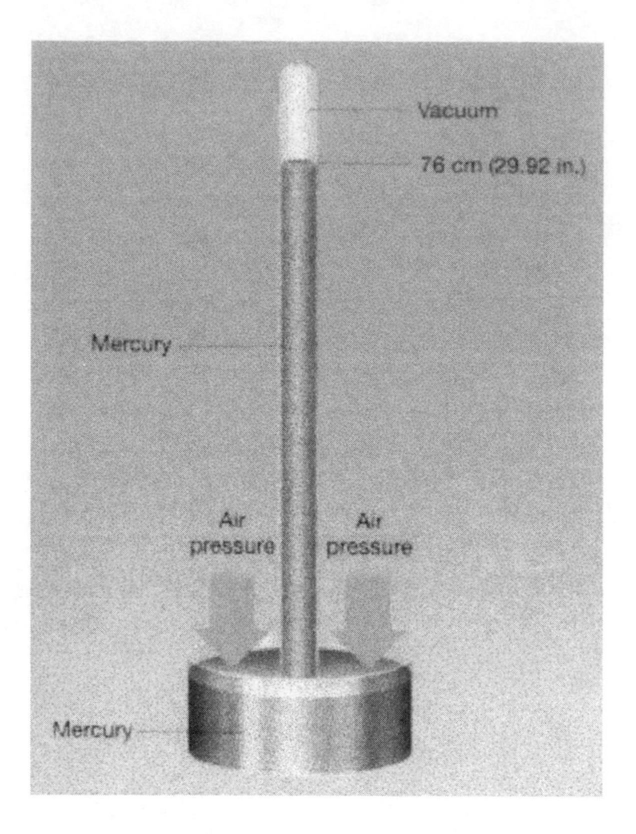

Fig. 5.9
Torricelli's Pressure Demonstration
(*The Atmosphere*: 2001-159)

Those expectations over a repeated cycle that are covered by a physical law are the behavior we seek.

The standard weight of air pressure at sea level is 14.7 pounds per square inch. This represents a column of atmospheric molecules in an area of one square inch from sea level to the top of the atmosphere. This measurement is an average amount. If more molecules became attached to those in the average measurement, than the pressure of 14.7 pounds would increase, and if molecules within the column of atmosphere migrate out, then there are fewer of them and the pressure of 14.7 pounds will be correspondingly lower. Because terrestrial topography varies, the average air pressure will also vary due to altitude differences and point of measurement. In addition, since the air pressure will vary and exert a different surface pressure, other liquids or solids would also experience that force change will have demonstrated it accordingly.

In 1643, one of Galileo's students was an individual by the name of Torricelli. He identified this pressure and for the first time postulated that the amount of air around us pressed upon all things. To demonstrate his thinking, he closed off one end of a glass tube, inverted it, and placed the open end into a dish of mercury. He allowed the mercury in the tube to flow out until the mercury column corresponded to a specific height within the glass tube (fig. 5.9). Torrecelli observed that at certain times the mercury level within the tube rose and fell. His conclusions was that as the air pressure pushed downward on the dish surface, the corresponding displaced liquid would rise within the tube. He also concluded that as the air pressure decreased, the column within the tube would possess a greater weight. That would lead to a consequence of flowing out of the tube until the column's weight was again equal with the downward dish surface's pressure force, which was a balance against the vacuum at the top of the glass tube over the mercury. This measuring of that column of mercury is known today as "inches of mercury" (Lutgens and Tarbuck, 2001:159). The level of the average mercury that Torricelli measured was 29.92 inches. For standard use, the U.S. Weather Service has adopted the term for a standard column of mercury as *millibars* or *100 newtons* (which is the physics term for atmospheric pressure) per square meter or 1013.25 millibars for average sea level pressure.

Readings in air pressure of millibars or inches of mercury vary globally, and as the measurement points are taken in these locations they confirm that the possibility exists for differences. Because multiple points of measurement have variations and occur at the same moment in time, then the differences of millibaric pressure indicate that a variable is present to create that possibility. That variable is the influenced reaction of the molecule and atom to temperature, called thermodynamics. As the atmosphere experiences a lessening of molecules with height and the temperature of a gas is related to the number of collisions its internal molecules contain, that variation will promote a temperature change relative to a specific point within the gas area. Because a decrease in molecules with height is to be expected, and with height comes a variation of temperature called the lapse rate, an idealized table of those expectations in figure 5.10 demonstrates the resultant pressure with those factors.

In the previous study temperature, pressure, and relative densities all were in support of a perfect gas. Our atmosphere is also considered a per-

Height (KM)	Pressure (MB)	Temperature (°C)
50.0	0.798	–2
40.0	2.87	–22
35.0	5.75	–36
30.0	11.97	–46
25.0	25.49	–51
20.0	55.29	–56
18.0	75.65	–56
16.0	103.5	–56
14.0	141.7	–56
12.0	194.0	–56
10.0	265.0	–50
9.0	308.0	–43
8.0	356.5	–37
7.0	411.0	–30
6.0	472.2	–24
5.0	540.4	–17
4.0	616.6	–11
3.5	657.8	–8
3.0	701.2	–4
2.5	746.9	–1
2.0	795.0	2
1.5	845.6	5
1.0	898.8	9
0.5	954.6	12
0	1013.2	15

Fig. 5.10
U.S. Standard Atmospheric Expectations
(The Atmosphere:2001-163)

fect gas, which contains water molecules that encompass a volume known as dry air. In this case the water is a gas we call water vapor and will behave accordingly to the same law as any other gas. We know that water may also be in either a liquid or solid form depending on the temperature to which it is exposed.

If water is injected into a specific volume, it will evaporate and exert a certain pressure in that state. As more water is injected, a point will be reached whereby evaporation will no longer take place and the exerting

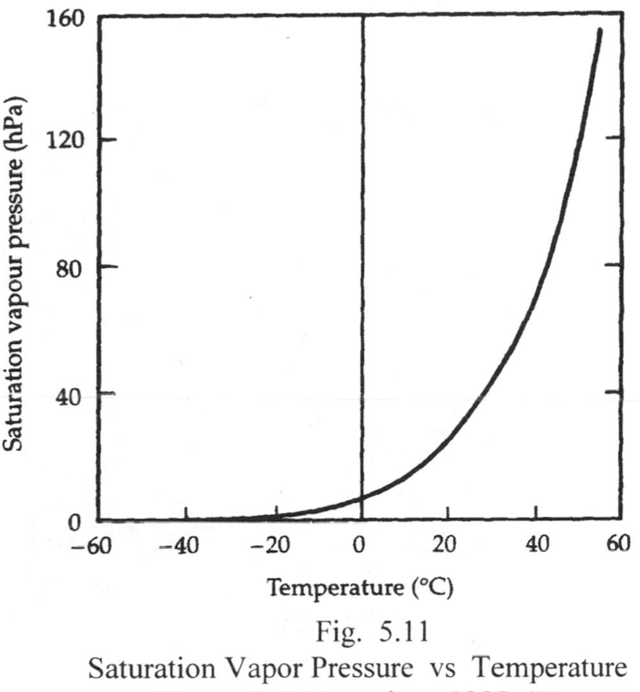

Fig. 5.11
Saturation Vapor Pressure vs Temperature
(Dynamic Meteorology:1998 41)

pressure will remain at a constant, indicating that the point of saturation pressure has been reached (Gordon, Grace, Schwerdtfeger, and Scott, 1998:32). This point is dependent on temperature only and will increase as the temperature increases (fig. 5.11).

Although other gases are present within the atmosphere, our concentration within the molecular realm will be on the investigation of water vapor as a constituent and the subsequent definition concerning that principal condition as moist air, along with an analysis of how much moist air is present and any conditions that are a result of its behaviors. At the saturation pressure, there is an implication of abundance and that the abundance will follow Dalton's Rule, also that the atmosphere is homogeneous, that is, having no clouds, and that the total pressure of the system is equal to the sum of all pressures represented by the gas constituents (1992:52).

As a matter of fact, moist air is less dense than dry air and at a moist

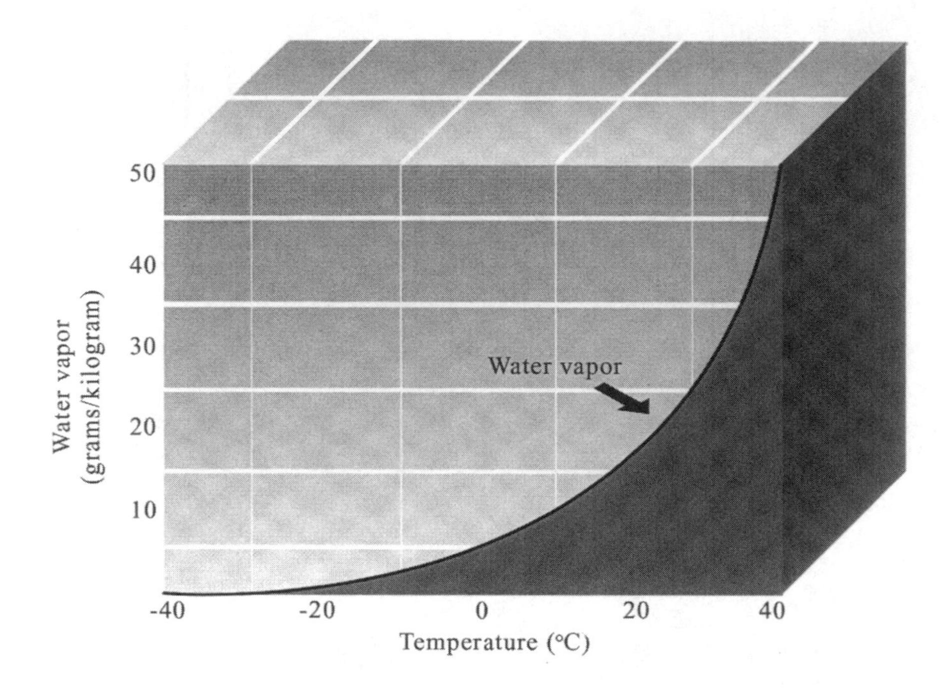

Fig. 5.12
Water to Saturate 1 kilogram of Dry Air at Temp.
Variation from text
(Mendes-Hussey Graphics)

air pressure its virtual temperature will always be greater than that of a dry air system. The virtual temperature is that temperature to which dry air must be elevated to contain the same density that moist air would have at the same pressure, which explains why it takes more water vapor at a higher temperature to reach saturation levels illustrated by figure 5.12. The content of moist air is important to many facets of life, and because of that the understanding between general atmosphere totalities representing all constituents and the relationship to pressure and temperature can be analyzed with what has been covered so far.

Vapor Ratios

Study of the moisture content of air is very complex, because it is expressed by ratios. It is more of a capacity to hold or retain as an amount of the total material involved. If a specific pressure and temperature are selected, that is, at a specific millibar or inch of mercury at a specific temperature in degrees Fahrenheit or Celsius, than the amount of actual water vapor in existence, compared to the amount of water vapor that would be required in the atmosphere to reach the saturation level, is called relative humidity. This number will in effect indicate how close to the point of water vapor saturation at that pressure and temperature the circumstances are, represented by percent and not how much water vapor in quantity exists, although content of water vapor is part of the basis for the conclusion. Because relative humidity is a mathematical value, then it is subject to change if one of the components does. Since relative humidity is a percentage or ratiomatic, the end calculation will be changed if the content of water vapor is either less or more. Also, since the water vapor saturation point can be changed by a change in air temperature, the pressure will also change, as it is dependent upon that. These lead to the conclusion that relative humidity will vary depending upon air temperature (Lutgens and Tarbuck, 2001:94).

The general amount of water vapor that may be in the air is not necessarily a ratiomatic value as much as a general term of content. It has several expressions, one of which has been discussed in the previous section, that is, relative humidity. Another expression of humidity may be absolute humidity. The term *absolute humidity* refers to the mass of water vapor that exists within a specific volume of air and is expressed usually in grams per cubic meter. As air is in motion, both pressure and temperature will affect its volume and with this variance the value of absolute humidity will change.

The third way of expressing the amount of water vapor in the air is by the mixing ratio. Mixing implies more than one, and the atmosphere consists of water vapor and the dry air component of N_2 78 percent) and O_2 (21 percent), with a balance of trace gases considered to be a molar fraction (1996:7) (fig. 5.18). To determine the moisture content of the atmosphere

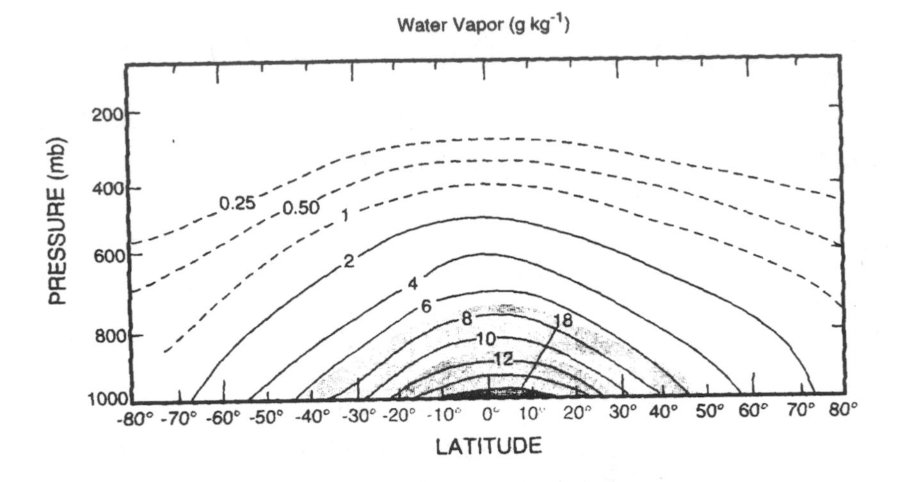

Fig. 5.13
Zonal Mean Mixing Ratios vs Pressure and Latitude
(Atmospheric Physics:1996-26)

within specifics of humidity, direct sampling may not produce the specifics required, so a measurement of vapor pressure (additional molecules through evaporation), relative humidity (ratio percentage), and the dew point are used. The dew point is a temperature value that represents at what point a specific volume of air would need to be cooled to in order to reach saturation. Should the air temperature within the specific volume of air cool below that saturation point, than the excess water vapor as a gas will condense out and be visible. The dew point illustrates that objects near the ground will be cooler if the air volume is at the saturation point and then the objects at ground level will be cooler and formations of water will form on them because of condensation. Other forms of visibly condensed air may be in cloud forms such as fog. Because dew point saturation (not condensation) is temperature-dependent and moist air requires a higher temperature that dry air, a high dew point would indicate moist air and a lower dew point would be required to indicate dryer air. The compositions of the mixing ratio of water vapor (contour) and density of water vapor or absolute humidity (shaded) as a function of pressure and temporal locale are illustrated to identify this study in figure 5.13.

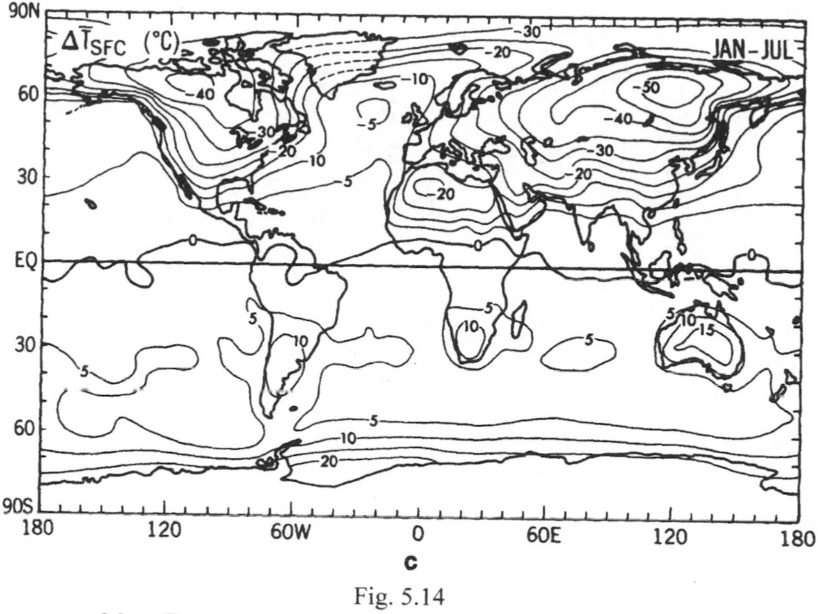

Fig. 5.14
Mean Temperature Difference, Jan. and July 1963-1993
(Physics of Climate:1992-139)

Temporally speaking, the Earth has many features that will control to some extent how much water vapor will be in the atmosphere at any given time. Since the atmospheric capacity to retain water vapor is temperature-dependent, an analysis of the mean surface temperatures should be examined along with the moisture content. The illustration of figure 5.14 details the surface distributions in degrees Celsius for the difference between January and July for the period of 1963–73 (1992:139). With this valuable information, a comparison of the mean zonal variations of humidity can be better understood. As expected, the highest values of humidity, eighteen and nineteen grams per kilogram of volume of moist air, are found over tropical areas (fig. 5.15). The presence of oceanic cooler and warmer water will identify the deflection that is apparent near the eastern and western coastlines.

Because water vapor is concentrated within the troposphere, more specifically within the lowest two kilometers of the atmosphere, temporal activity of condensation and evaporation locations are of importance. Spe-

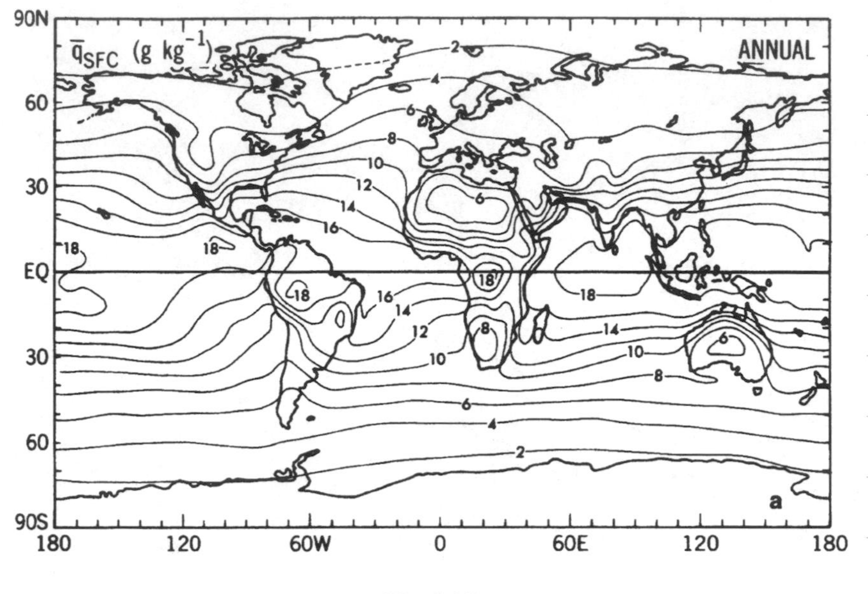

Fig. 5.15

Mean Zonal Distribution of Relative Humidity
(Physics of Climate:1992-279)

cifically, the condensation into a manifestation that is directly observable and experienced by man is the morphology of vapor and the results from it.

Aerosol Morphology

For the atmosphere to hold water vapor it must sustain suspensions of a variety of molecular sizes. The suspensions of various-size molecules and particles are necessary to induce cloud formations, which without the suspended particles they would most likely not form. Here, however, we are going to consider the formation of a postcondensation situation in the droplet. This is a case in which the particular growth of a droplet is through homogeneous nucleation. In this process, only water vapor will be considered and not a growth process that would require a nucleus of some aerosol particulate.

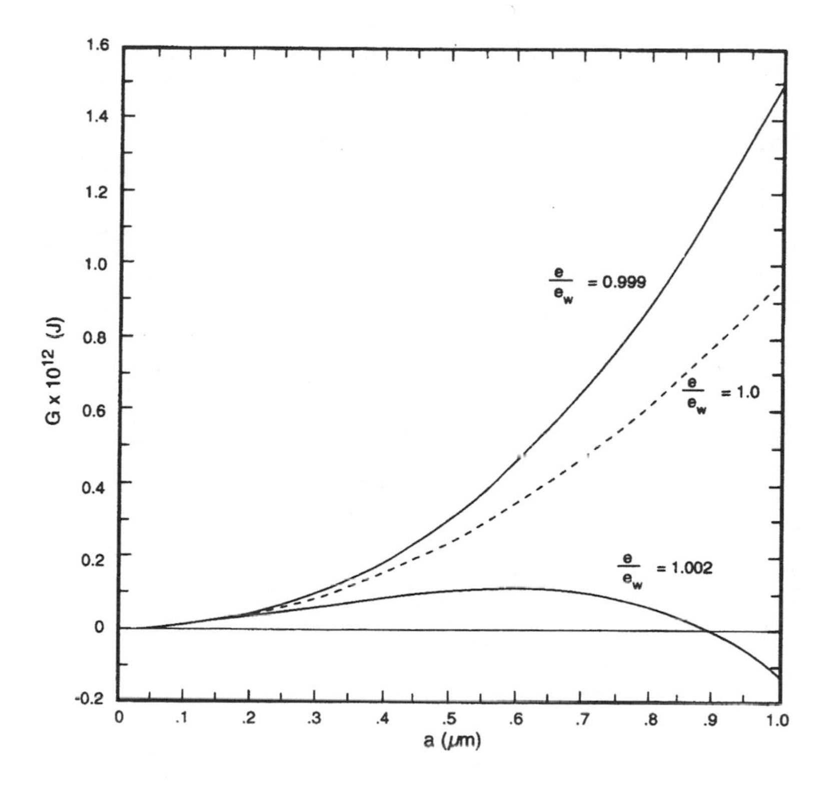

Fig. 5.16
Pure Water Droplet from Homogeneous Nucleation
(Fundamentals of Atmospheric Physics:1996-267)
Function of droplet Radii to: subsaturation, saturation and supersaturation
note: e/ew = <1, =1, >1

The survival of a small embryonic droplet that would occur from a chance collision with water vapor molecules depends upon the delicate balance between condensation and evaporation. It is the survival time that is predominant here, due to the fact that it must be in existence long enough to increase its size. The state of maintaining an equilibrium balance is considered a process of work by the droplet and is a function between its surface area and the energy that such a surface area will produce. This is called free energy and is a mathematical function of the gas law for vapor, Boltzmann constant, the radius of the droplet, and saturation pressure illustrated by figure 5.16 (1966:267).

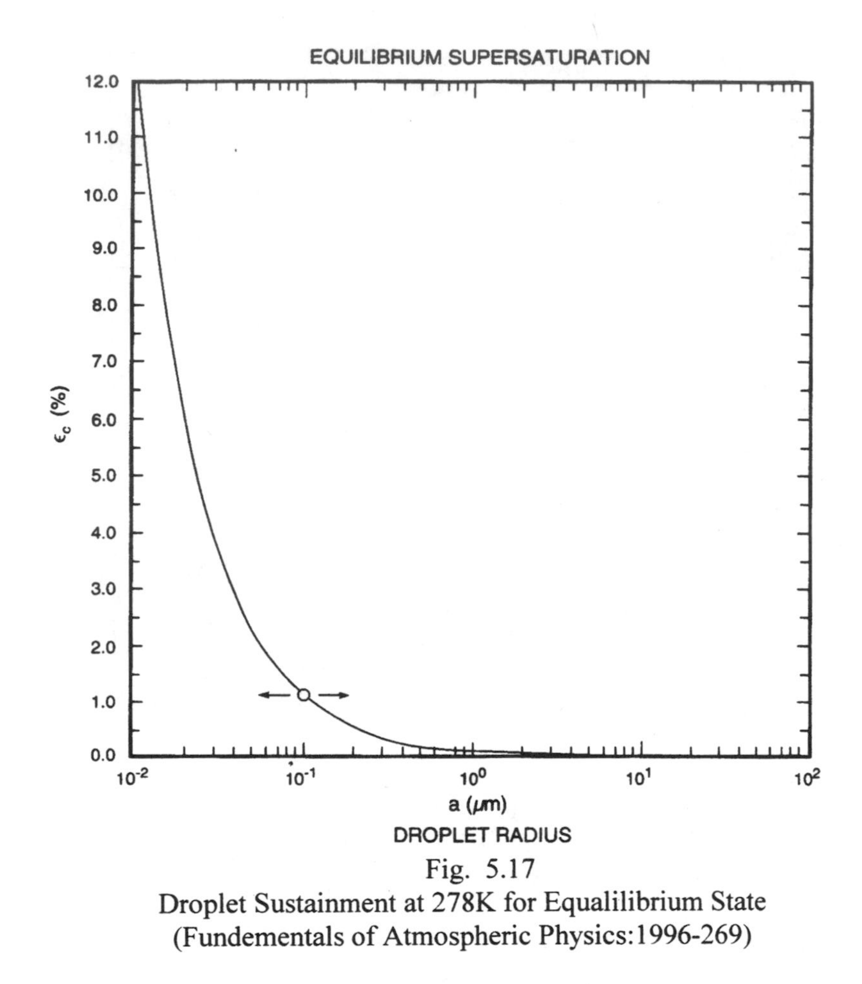

Fig. 5.17
Droplet Sustainment at 278K for Equalilibrium State
(Fundementals of Atmospheric Physics:1996-269)

For a drop to exist in an evaporative condition, that is, to maintain its equilibrium, replacement of molecules from its surface is necessary or the droplet itself will eventually disintegrate. The rate of evaporation is dependent on the surface curvature of the droplet, by which the greater the curvature, that being a relation to droplet size, the greater the rate of evaporation. This process is called the curvature effect, and because of this the smaller the droplet, the greater the vapor must be to prohibit evaporative tendencies. Since air may be saturated with respect to a flat surface, it may not reach the saturation point where a curvature is involved. Because of this, air around a small droplet must be above the saturation point of 100

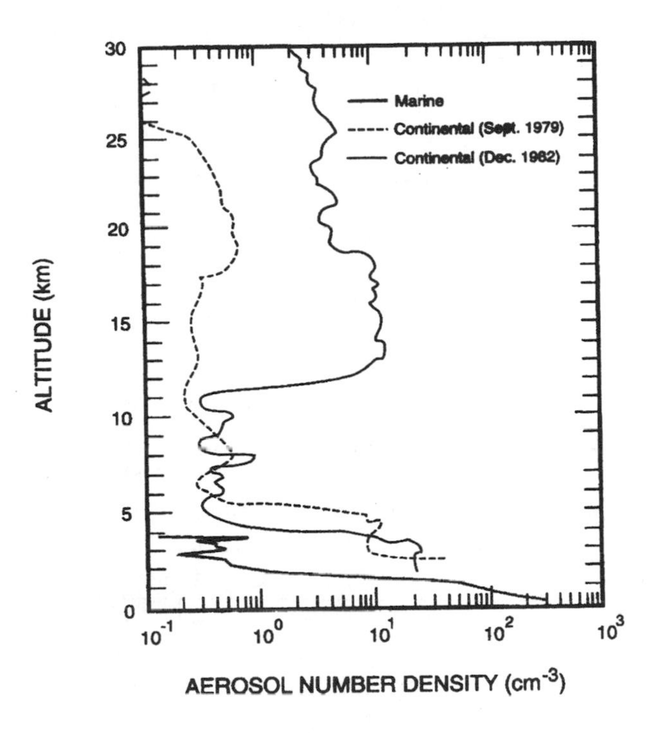

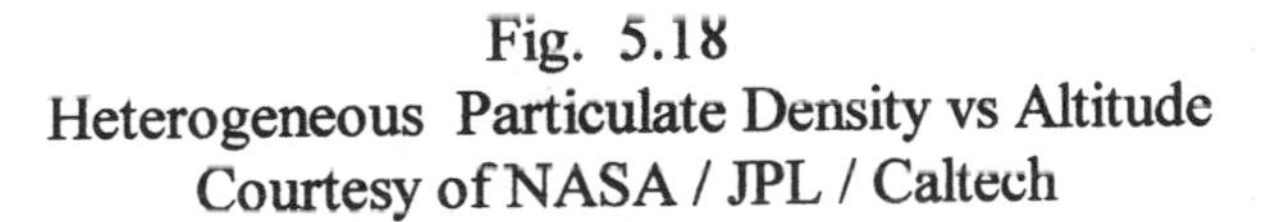

Fig. 5.18
Heterogeneous Particulate Density vs Altitude
Courtesy of NASA / JPL / Caltech

percent to keep that droplet in equilibrium; further, the smaller the droplet, the higher the saturation must be, i.e., supersaturating. With this thinking, the small curved surfaces of less than two micrometers will require supersaturation, above ten micrometers the surface radius is considered flat, and the supersaturation (percentage of relative humidity) requirement is not needed for equilibrium (fig. 5.17).

The droplet's falling through the atmosphere is an influence on its growth along with its radius. These two parameters are very important in size progressions and show that predictabilities can exist giving evidence that radius size is an acceleration to growth (1996:274) (fig. 5.18). As far as falling velocity parameters are concerned, droplets that are smaller than twenty micrometers in diameter will grow primarily through the condensa-

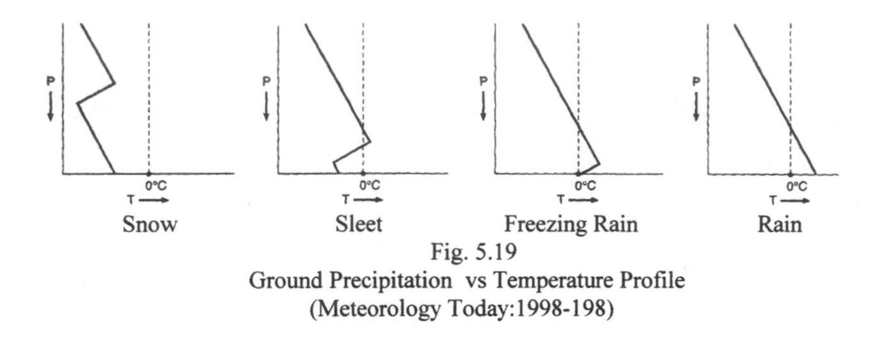

Fig. 5.19
Ground Precipitation vs Temperature Profile
(Meteorology Today:1998-198)

tion process, because their velocity would be about the same as the surrounding air and that inhibits collision tendencies. Greater than twenty micrometers will promote collisions and because collisions are expected more because of the greater size, the spectrum of growth will be seen to favor this population. When a droplet reaches 100 micrometers than its dominance would be viewed as precipitation at ground level (1996:274).

Pure water molecules at lower successive temperatures react in a way that can be seen as crystalline structures, and the term *homogeneous freezing* may be used to define it. If ice-forming structures called ice nuclei are present and become active through lower temperatures, then ice crystals may form; however, there must be enough water molecules that will combine together in an ice embryo to cause crystal inception. An ice nucleus may be in the form of clay minerals, decaying plant bacteria, or ice crystals themselves. In nature there is a rarity of lattice structures that mimic those of ice, so the condensation process is anticipated to be the primary method. Ice nuclei that allow water vapor to affix directly to them in cold saturated air are termed *deposition nuclei.* In this situation, the water vapor molecules will form ice directly without first going through the liquid phase and enhance additional formations in cold air. These formations can be considered supercooled and termed *freezing nuclei* (Ahrens, 1994:196). Of coarse nuclei that come into contact with others through the collision process are defined as contact nuclei, and the process that is formed from the collisions is called contact freezing, which some studies have concluded may be almost any nuclei and the primary cause of ice crystal production.

Because the saturation vapor pressure is greater over the surface of water than that of ice at the same subfreezing temperature, more water valor molecules will surround a supercooled water droplet than an ice crystal. It is the vapor pressure difference that causes the diffusion from a water drop to be attracted to an ice crystal. Since the water molecules are now less in atmospheric quantity, a lower vapor pressure will exist over the droplet and this imbalance will cause the water droplet to evaporate and replenish the water vapor removed by the ice crystals. It is for this reason that ice crystals have a ready supply of vapor to grow by absorption and through this deposition process grow at a rapid rate. As the ice crystals form and come into contact with each other, called aggregation, some may fracture into smaller structures and adhere to other crystals also in free fall, in so doing, a pattern will form that we identify as a snowflake (Ahrens, 1994:197).

In conditions where the snowflake will melt into a water droplet and again into another morphological condition, several occurrences must take place in the flake's path toward the ground. The temperature profile is an important factor. With the snowflake, the temperature from altitude to ground level was below zero degrees Celsius. Should the temperature profile be higher at the ground than at altitude, rain would be expected. If the temperature profile passes through and above zero degrees Celsius as the flake falls to ground level at some lower altitude, the flake will melt at that point. If the temperatures profile then returns to a point below zero degrees Celsius prior to ground level, the droplet at that transition point will again freeze into ice, not a snowflake, and we term the ice pellet prior to reaching ground level as being *sleet.* Also, if the temperature profile is such that the transition point from below zero degrees Celsius to above zero degrees Celsius is at some point still at altitude again the flake will melt into a droplet, and if the temperature profile then falls below zero degrees Celsius at ground level, then the droplet will fall to the ground and freeze in contact with objects on the ground, in this case freezing rain (fig. 5.19).

Other vapor observations related to our atmosphere as a function of a water molecule study can be observed through its ability to form molecular adhesion with a particle nucleus other than itself. Since droplets may increase their size through condensation, they would require additional water vapor to be present in sufficient quantities to sustain formation and this process being the simplest of nucleation. In addition, even though the droplet may be as large as 0.01 micrometer, sustainablilty requires 12 percent supersaturation; this is important when greater than .1 percent supersaturations are rare (1996:268). It is therefore a conclusion here that what is a visible representation of water molecule densities must have other than a homogeneous water vapor atmosphere to begin embryonic nucleation. We know that our atmosphere consists of nitrogen, oxygen, and trace elements. We also know that an ice crystal will be formed around other nuclei that comprise additional aerosols, so this path is the most plausible solution.

Accepting that the atmosphere around us consists of particulate that would identify the anthroprogenetic character of our existence in aerosol matter, suspended of course to the point of molecular interactivities with a water molecule, that define a heterogeneous nucleation process (1996:268) and similar to crystal embryonic nucleation discussed earlier. Within the heterogeneous aerosol are small particulates of sodium chloride (NaCl), NH_4, and ammonium sulfate (SO_4) that demonstrate the nonhydrological purities with additional nitrates, suspended as a byproduct of combustion. Once these particles enter the tropospheric layer, they will meet and collide with water vapor content (humidity), upon which, a reaction with the water molecule takes place. Although altitude is a factor, the reference to figure 5.18 will illustrate the densities as a function of it. The absorption of water by these nuclei defines them as being hygroscopic, and because the subject is concerned with a pure substance, that being the water molecule, than the nucleation process is considered to be heterogeneous-homomolecular (Seinfeld and Pandis, 1998:545). Although ionic nucleation does occur, that is, water molecules and ion particles will nucleate, the information here is presented with the aerosol particle only.

As molecules collide with each other, there is a thermodynamic exchange in the form of heat; this is a short time frame in comparison to the time scale of both reduction and addition of water molecules. For this reason, a thermoequilibrium will exist with the molecules sooner than equilibrium with the density of their surrounding vapor. An equilibrium with the molecular cluster that is advancing in growth will then be greater than the change in the atmospheric concentration of molecules (saturation). At this stage a visible observation may be made. That visible observation is the cloud.

6

When Molecules Gather

Ancient Observation

The water molecule will float above us in vast quantities. Is there observable evidence of this? What does that quantity represent and how are their formations related to pressure, temperature, and height? Can they be observed through means other than eyesight? Are they the evidence that we seek to expose reactions that define atmosphere and when did the observations develop that provided science with such confirmation?

In the earliest of times, man saw unique manifestations of the water molecule in vast densities. He didn't know what he saw, only that it was and then was not. As centuries progressed, man grew to understand that whatever these formations were, they did something to the area when they showed up and sometimes their presence meant good and sometimes their presence meant hardship. They hid the sun by the very nature of their passing, and their passing sometimes was near and sometimes distant. However great the distance, they caused the sun to go away.

At first man believed that the phenomena witnessed were the rationale of a higher order. In the centuries of the early Chinese, there was a god of thunder and lightning, Master of the Rain, and so forth (Hamblin, 2001:24). Although the reasoning behind man's conception of authorship to these formations may have been rooted in cultural belief premised on the divine state, the formations themselves had a design and characteristic that seemed to make them more important from the reality standpoint.

As people migrated over landmasses that now are considered to be North America, South America, Europe, Asia, et cetera, topographical locales shifted. With this shift in topography, observations of these formations became indicators of weather patterns that were never experienced. Because cultures could not adapt in preparation for changes that would have devastating consequences for them, many perished and did not know why. In some cases, months of travel to a seemingly pleasant place having different geological features may bring subzero temperatures rapidly, only to persist for long periods of time. But the formations in the sky were different from where the travelers originated, and the pattern of their arriving and going was different also. It is here that events concerning the formations passing overhead caused the idea that, perhaps their passing should be recorded for forecasting. In the fifth century B.C., the Greeks attempted to do exactly that, to such an extent as to place in writing on the columns in the public squares within cities about the Mediterranean observations regarding cloud formations and local conditions (Hamblin, 2001:32). Without a more precise analysis concerning the formations, the Greeks could only speculate on what might occur without confidence, and although the Atomists had really presented the final piece to the phenomena, with their proof floating above them, the two were never linked.

Not until the seventeenth century when Rene Descartes (1596–1650) published in his discourse *Discourse on Method* (1637) the insight that the formations visible were a physical reality that was founded in the philosophies held by the Atomists. He presented the idea that it was from these formations water fell and the mechanism for that process was the fact that small parts of compressed ice or water vapor from the ground became larger through a growth process. Although this has exceptional merit, it was nevertheless not in the forefront of thinking, and in the middle to late 1700s formations were still considered to be some vaporous substance, filled by rarefied air and heated by the sun to expansion until they burst, whereby the contained water is released (Hamblin, 2001:45). Finally a change occurred that provided the impetus to focus on the air and would eventually produce a comprehensive catalog of molecular formations that become visible and adhere to certain principles.

In 1783, London experienced a significant meteorological condition that became most severe. It was this severity that caused the scientific community to give more attention to the air around them. Hot days and very cold nights in association with the sulfur from their living habits created a situation that became most toxic. It was during these times that a student named Luke Howard would remember and apply to his eventual paper, submitted before the Askesian Society as "On the Modification of Clouds, and on the Principles of Their Production, Suspension, and Destruction; Being the Substance of an Essay" (1802:3; Hamblin, 2001:171). It was this essay that termed *cloud* as a morphological presence, complete with classification and definitions consistent with Simple Modifications, Intermediate Modifications, and Compound Modifications, seven in total. Such became the first steps of a categorical and historical study of clouds, today termed *nephology*.

Luke Howard's premise for the paper was presented to describe physical bodies consisting of suspended water and the changes those suspensions will develop into through compression. In addition, the changes are a physical result of internal motion. Later in the centuries that followed, a more detailed analysis was developed to enhance the reality that clouds are a representation of atmospheric and planetary thermodynamics, of which will be covered in some detail later.

Clouds have different shapes and sizes and will develop into formations that indicate specific atmospheric tendencies. By that I mean rain, sleet, hail, fog, and a beautiful weather period in either advancement or retreat. Clouds will also indicate the relationships between altitude, temperature, and pressure by their densities and general sequence over a period of time. Using Howard's original cloud identification, in 1887 the Honorable Ralph Abercromby of Belgrave Square and Prof. H. Hildebrand of the University Observatory of Uppsala, Sweden, felt that although Howard's classifications were the best available, the experiences of worldly travel by Abercromby and his documentation of the patterns witnessed at a variety of geolocales would enhance the already-instrumental work. Both Abercromby and Hildebrand in 1890 published the findings complete with photographs and updated the original seven classifications to ten (Hamblin, 2001:341), continuing to use Howard's terminology. Clouds were now classified by the altitude at which they could be found and additional information to the structural formation delineating rotund or sky covering was also included. Today the clouds that we can see passing overhead now have a place in science and history. What was unnamed and unknown is now the cornerstone of meteorological communities, complete with self-identifiable characteristics and language.

Clouds

Clouds are great; they are fascinating to watch and fascinating to understand. They bring planetary cooling and water over landmasses and provide shade. Now that they have been classified and analyzed as forma-

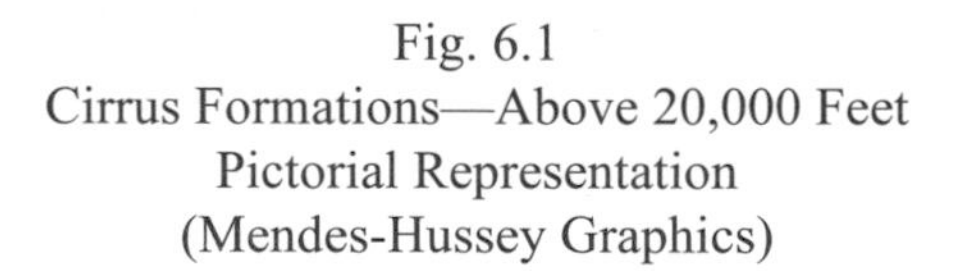

Fig. 6.1
Cirrus Formations—Above 20,000 Feet
Pictorial Representation
(Mendes-Hussey Graphics)

tions of visible aggregates of minute droplets of water or crystals (Lutgens and Tarbuck, 2001:120), learning to understand them can be a most worthwhile study.

From the highest cloud developments come the form of cirrus (*cirriform:* meaning "fibrous") or mares' tails, the cirrostratus, and the cirrocumulus. They look like very wispy streamers and usually form at altitudes above 20,000 feet, which is the upper boundary of the troposphere (fig. 6.1). At this altitude, the temperature is so cold that any moisture within the air envelope is condensed out as ice crystals. The microphysical aspect of cirrus clouds is in the particle size, as all particle sizes will vary with formations. Particle number densities for the cirrus formation will vary from 10^{-7} to 10^{-3} centimeters with crystal dimensions in the range of

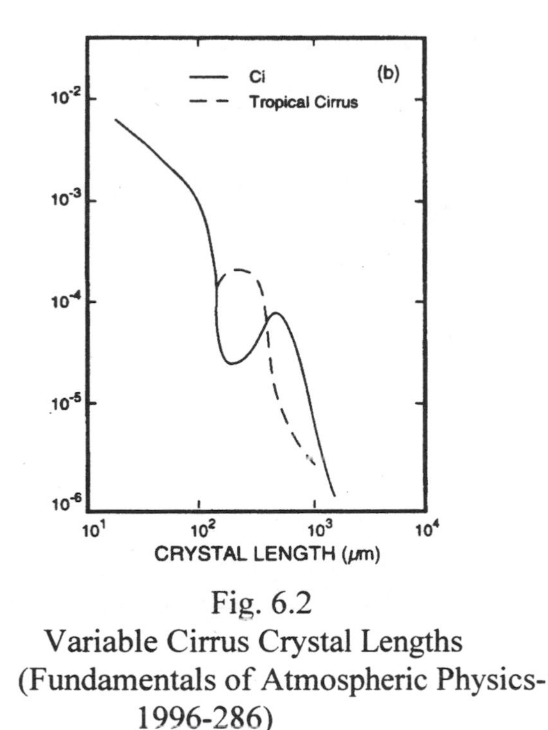

Fig. 6.2
Variable Cirrus Crystal Lengths
(Fundamentals of Atmospheric Physics-
1996-286)

as small as 10 microns to a large 1 millimeter (1996:286). It is because of the larger crystal sizes that will promote the observed tail filament characteristic, which is due to rapid falling crystal velocities (fig. 6.2). Although the higher clouds such as cirrus will not cause precipitation, they offer a forecast that lower altitude and greater density cloud formations will follow.

Cirrostratus clouds are similar to the mares' tails structure except that these clouds generally cover the sky to a greater extent and seem to make the sun appear as though it is shining through a murky glass, giving an opaqueness quality to the light. In view of this effect, the light will have a central glow that is identified by a halo or circle of refracted light surrounding the sun or moon. The stratus definition identifies a layered effect that will cover most or all of the sky; the cirrus defines the crystalline structures that comprises the density and because of the very nature of the crystal produce the halo effect (fig. 6.3). Because cirrostratus clouds are generally followed by clouds at lesser altitudes, they can be harbingers of moisture within twenty-four hours.

Fig. 6.3
Cirrostratus Formation
(*The Atmosphere*: 2001-123)

The last of the three types of high-altitude clouds are the cirrocumulus. These clouds generally can be identified by formations that produce small, isolated cotton ball effects that may be seen in a rows characteristic (fig. 6.4). The puff effect is because of an upswelling of air that will cause the moisture to ascend vertically and in a buoyant manner. The cumuloform aspect is derived from heat both as a result of surface transfers and as latent heat that is a product of the internal condensation process. Because of the vertical nature of thermal growth into higher regions of lower temperature, these formations may last from minutes to hours. The buoyancy is not sustained when the outer environmental air is drawn into the cloud's center; this is a displacement and will destroy the buoyant effect by the mixing of the internal air. At this point the central core of the cumulus structure is transformed into a toroid by the thermal expansiveness of the cloud, it becomes removed, and the cloud's buoyancy status dissipates (1996:277). At lower altitudes, the cumulus effect may be sustained due to the protection the core will have by the stability of the surrounding intrusive saturated air. The distinguishing effect of this high-level formation is

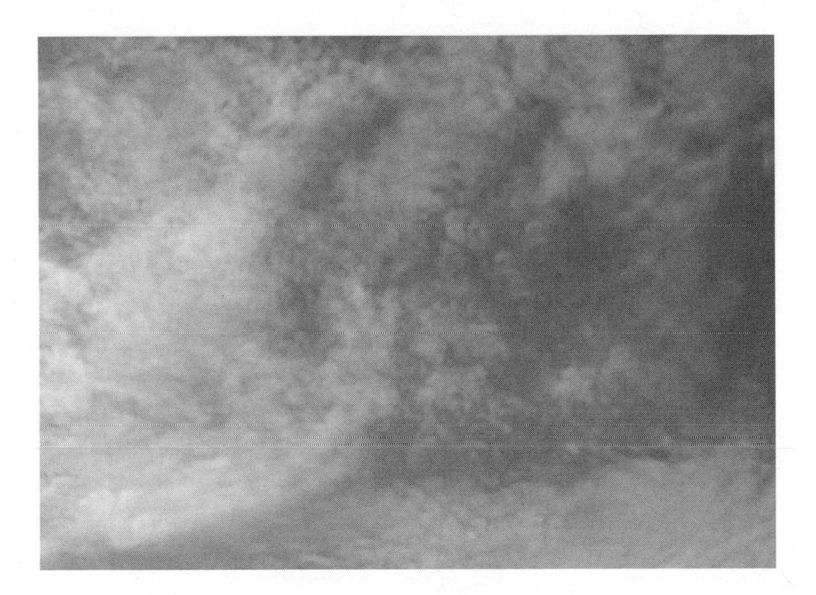

Fig. 6.4
Cirrocumulus Formation
(Interpretation of Text)
Mendes-Hussey Graphics

the ripple definition that differentiates it from the sheetlike cirrus, lending itself to the identification of mackerel sky or fish scale. Because lower clouds tend to contain more water, the color would be darker than that of the higher cloud formations that contain a greater preponderance of ice crystals.

The next type of clouds belong to the middle atmosphere or between 6,500 and 23,000 feet and are referred to with the identifier *alto.* Since cloud tops are in some cases very high, the composition would be that of ice crystal and supercooled nuclei; the lower portions would contain water droplets instead of the higher altitudinal composition of ice crystals because the temperatures at the clouds' lower portion would certainly be well above zero degrees Celsius, which would be necessary to achieve the liquid phase.

Because of their middle-altitude locations, the circulative aspects of the atmosphere may be seen in them as a representation of eventual thunderstorms and weather frontal systems. Two formations that can be observed in the middle atmosphere are those of the altocumulus (fig. 6.5) and

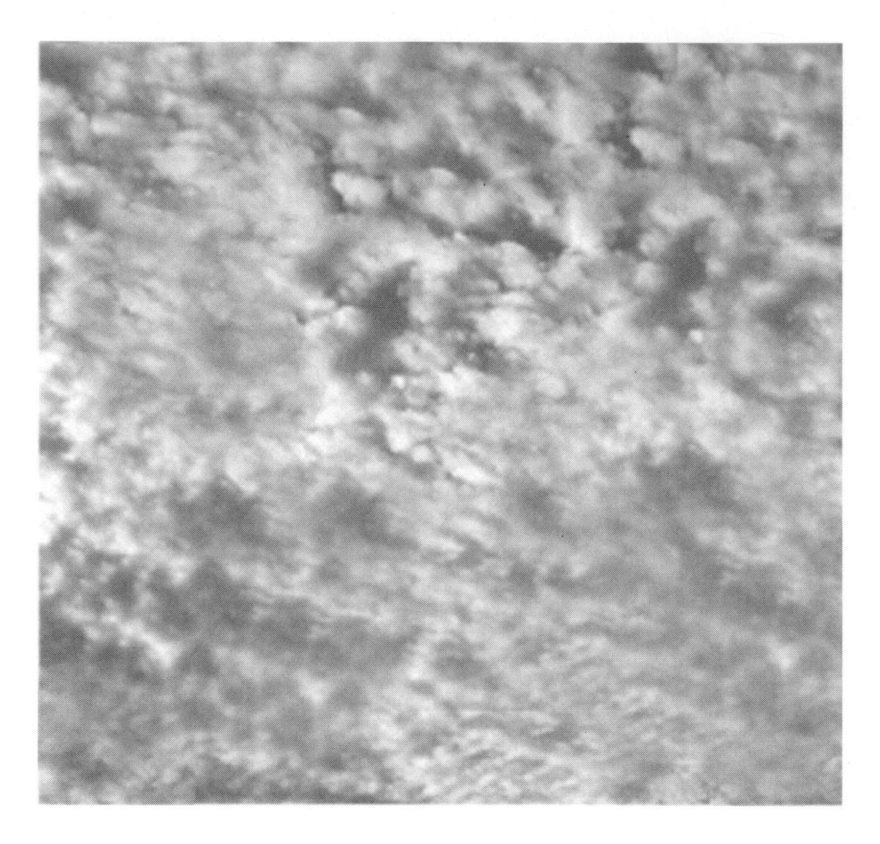

Fig. 6.5
Middle Clouds
Alto Cumulus Formation
(Interpretation of Text)
Mendes-Hussey Graphics

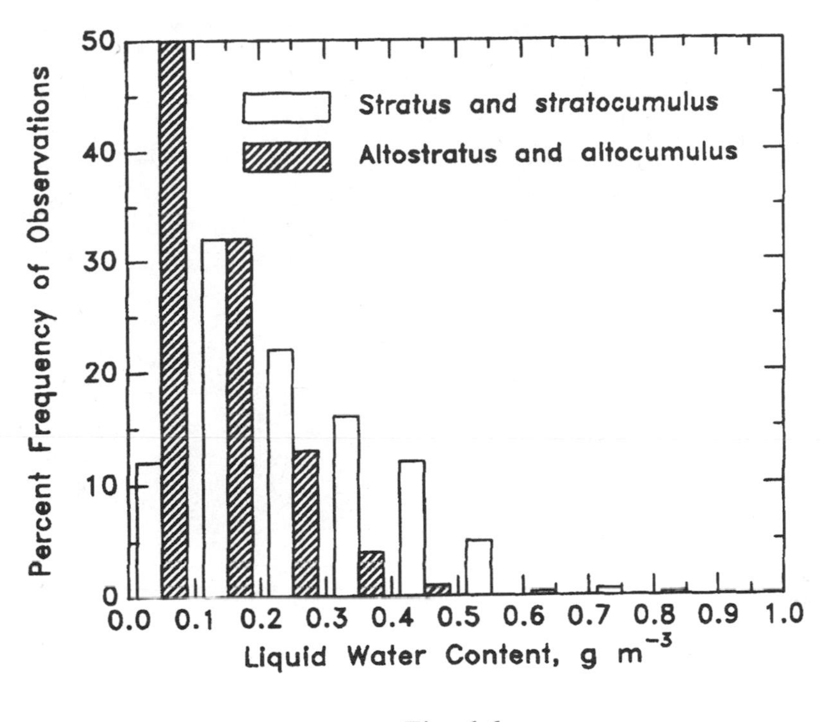

Fig. 6.6

Frequency Distribution of Cloud types over Europe
(Atmospheric Chemistry and Physics:1998-340)

altostratus. Due to the stratiform nature of the formation, a typical image of altocumulus will look quite similar to figure 6.3, only thicker.

Altocumulus structures are composed of clumps that will be arranged in a patchy display, much resembling that of the back of a sheep in a wool-like pattern, with areas of blue sky in between the clusters (Lutgens and Tarbuck, 2001:124). This is due to the density of water composition rather than that of the structure of crystals and will correspondingly have an observable gray and white difference precisely because of that particular aqueous density. Since altocumulus structures resemble a similar pattern to the cirrocumulus, one method of identification to alto cumuli is to hold out your arm to its full length and extend only your thumb pointed vertically. If the globular structures appear to be the size of your fingernail,

Fig. 6.7
Stratiform Cloud at Low Altitude
Interpretation from text

then the formation is most probably that of the middle layer of stratiform, or altocumulus. If the altocumulus clouds indicate a flattening top much like a small castle, then the probability exists that further vertical development into thunder cumulus structures may follow later that day (Ahrens, 1994:145), due to this particular representation designating rising air currents at lower altitude levels. It is these alto formations that are dominate in most water content, especially in Europe (fig. 6.6). Typically, these formations will be seen in a greater frequency over oceans, by 46 percent versus 35 percent over landmasses (Seinfeld and Pandis, 1998:339).

Altostratus clouds are very similar to cirrostratus in that because they represent the covering of an entire sky the *stratiform* designation applies. The *alto* identifier is due to the altitude at which this formation will be found. Aqueous densities of both water and ice crystals at low altitude will promote an opaqueness from the sun. The distinguishing feature of the sun's light being diffused is that due to the liquid portion of the total density absorbing light, the refracted portion is not available, thus preventing a halo, which cirroform formations can produce.

Fig. 6.8
Nimbostratus Formations
(*The Atmosphere*: 2001-126)

The closest clouds that form overhead will contain the stratus or flatfish formations and the vertical developmental formations of cumuli. Three total, two of stratiform and one of cumuli, will be discussed. The first type is stratiform. As previously considered, the stratiform will present a flat sky overhead with the coverage at nearly 100 percent. The base of these clouds will come very close to the Earth, and in areas where there are local places of height, such as mountain or city terrain, this is very evident (fig. 6.7). The bottom of the stratus would be smooth and not in a rumpled manner like a cumulus formation. The layered effect of the stratus and cirrus structure is indicative of a passage of either warm or cold systems and will be discussed as a representative of weather fronts later. Stratus formations will generally see greater occurrence while over oceans than over masses of land (Seinfeld and Pandis, 1998:339).

The thicker and denser water molecules within the condensation mass at a higher altitude and still considered as a lower cloud type are part of the nimbus structure. *Nimbus* is derived from the Latin word that means "rain cloud" (Lutgens and Tarbuck, 2001:124). Nimbus generally forms when the air is stable and will be caused by the convergence of other outside motion usually associated by a pressure gradient and forewarn of harsh weather. As the air intrudes and becomes vertical, the nimbus will become stratiform, developing a horizontal area that will be much greater than the layer's thickness. This atmospheric dynamic will result in the formation of nimbostratus (fig. 6.8) and possess greater water density than previously mentioned formations, forewarning of extended periods of wet weather.

Particles of a nimbus structure have a tendency to grow at an increased rate, and coupled with particle diameters of greater than twenty um (micrometers) the collision process is enhanced, thus promoting precipitave periods. Also, as a consequence of the stratiform nature of nimbus, the development for higher but weaker vertical growth is somewhat enhanced due to he absence of falling particle velocities affecting height growth. Because the primary destruction of cumuli comes from entrainment, within the stratifrom characteristic of nimbus, the process of entrainment, which removes the internal dominance of moist (saturated) air by mixing it with dry (unsaturated) intrusive air, is not promoted and this yields low, falling velocities so the positive buoyancy factor is diminished. Because nimbus clouds possess a lower base, terrestrial heating from below and a cooler top contribute highly to the disipative condition rather than the factor of entrainment (1996:286).

The next to last of the low clouds are stratocumulus. These clouds are similar to the altocumulus but larger. They will resemble a patchwork of white and gray with blue sky in between. As in the altocumulus, extending the arm and pointing your hand toward the clouds themselves, the size difference will be represented by your fist and not the thumbnail. Some stratocumulus may be the broken remains of larger cumulus formations and will generally be seen at or around sundown. Although they may grow into larger cumuli, these clusters normally do not issue precipitation. The stratus definition will describe relatively large sky coverage, and although the clouds themselves will be large, they will not possess great vertical growth. Notwithstanding, later discussions regarding the vertical and general motion dynamic will be discussed, and it is to be considered that in some cases where the vertical motion is halted and a more horizontal defining is observed the particles themselves contribute in many circumstances due to their weight and, will retard ascension. Also, atmospheric mixture, temperature, and radiation all affect the promotion of ascension. Internal particles that absorb radiation otherwise destined to reach the surface that encourages vertical convection will now also assist in that retardation (Davis, 1894:168). If the vertical ascension reaches an altitude at which the temperature gradient becomes warmer, then condensation is retarded and the forward motion will become dominant, with this situation the horizontal occurrence is advanced.

The last of the primary cloud images is the cumulonimbus. We have all seen them as a towering structure, extending vertically to immense heights. We have experienced their storms' being either rain or snow and intense nevertheless. They grow from the fair weather clouds of cumulus to cumulus congestus, resembling snow being pushed before a plow blade. From congestus the vertical growth is rapid until the tops reach approximately forty thousand feet, in which case the ice crystals contained within the top of the structure are displaced horizontally, presenting a flat appearance. It is this defining feature of the horizontal appearance that will distinguish the cumulus nimbus from the round defined top of a towering cumulus congestus (fig. 6.9). In some cases, cumulonimbus have been seen to extend over hundreds of miles, with the vertical structure ascending six to eight miles in height (Davis, 1894:169).

The previously mentioned clouds are the foundation of meteorological study and constitute a preponderance, additional clouds such as the lenticular formation, generally found over a mountainous area, developing in a layer somewhat resembling a UFO, and the banner clouds that form on the downslopes of mountainous regions are unusual and would be observed with less frequency.

The formation of a cloud has been discussed with the process of condensation of the water molecule about particulate nuclei and how that eventually develops into an observable phenomenon. Interestingly enough, the fact that cumuli can reach extreme upper atmospheric elevations is indicative of motion. Particles within the cloud structure are transported continually up inside, out, and back down, only to be drawn into a vertical convective situation at the base once more. This circulation is of course bringing saturated air up into regions that were unsaturated, only colder in

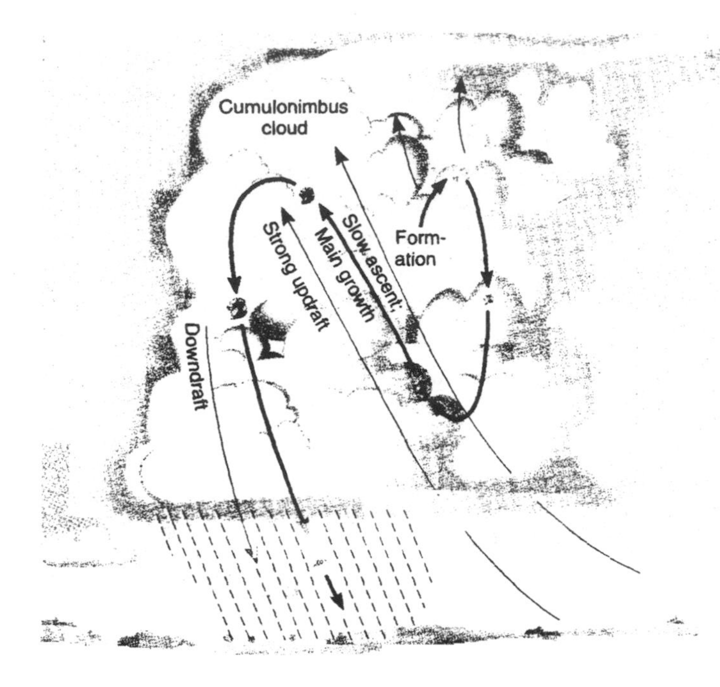

Fig. 6.9

Cumulonimbus Structure with Draft Regions

(Meteorology:1998-199)

temperature. The process is also indicative of something else, that a pressure difference exists between the cloud base and the top. This is partially due to the release of particle latent heat and will promote sustained vertical growth until unstable (unsaturated) air is reached where the temperature within the growth particle becomes warmer than that of its surroundings.

Previously discussed were definitions regarding saturated and unsaturated atmospheric conditions, all of which held something in common with temperature, pressure, and the relative humidity of water vapor. Under certain conditions, those combinations can be very favorable to us; however, at times when an imbalance of those conditions does occur, rendering large differences between any one of the three parameters, then an unstable atmospheric condition exists. To understand how the clouds represent these conditions of the atmosphere and the significance of height in concert with that, a further analysis of the water molecule and the environment of nitrogen and oxygen must be exposed.

We know that a water molecule is subject to a number of stimuli and will react in certain fashions as atmospheric behavior. Also, because the water molecule has a total atomic number that enables it to be buoyant with respect to the surrounding air—i.e., the molecular molecule atomic number is less than the atmospheric constituents of oxygen's and nitrogen's atomic numbers combined—suspensions of vast quantities are achieved. Clouds are suspended over a terrestrial surface because of this fact; however, this fact is insufficient when an explanation of the causes of horizontal water vapor motion and surrounding atmospheric constituents is addressed.

Clouds move; therefore droplet behavior and the associated condensation process are in transport en masse, as well as the atomic structures of nitrogen and oxygen. Taken further, there must be a transport mechanism that is in close association with respect to the surrounding nitrogen, oxygen, and water vapor environment and that the stability of that environment must be an influence to the mechanism.

Atmospheric Stabilities

The environment, we know, is comprised of atomic structures consisting of nitrogen, oxygen, and water vapor as the chief constituents. We also know that under certain conditions these elemental structures will behave in accordance with specific gas laws. And we also know that all gases respond to a pressure about them and in so doing affect the other parameters of temperature and density.

Close examination of a specific environment at the molecular level will provide great insight into the macroenvironment to which they belong. Meteoroligically, a hypothetical sample (size of large balloon) of our atmospheric environment is called an air parcel; it will consist of nitrogen molecules, oxygen molecules, and water vapor. Since a cloud represents some amount of water vapor and it is in either saturated or unsaturated air, its percent should be defined from a mass point of view as part of the total volume of the air within the parcel sample, including the other elements of nitrogen and oxygen.

Because vertical growth within a cloud can be established and the cloud is in motion, the air parcel will experience differences in the surrounding environment and will vary with topography and oceanic geolocales. In altitude, the air parcel will be subject to a lesser density of oxygen and nitrogen molecules, so the surrounding pressure exerted upon the air parcel's outward dimensions lessens; correspondingly, as the parcel moves closer to the surface, the surrounding air pressure will create a deformation response and contract.

The vertical dynamic causes an expansion of the air parcel's volume; however, the mass of water vapor does not vary. This volumetric increase while water masses remain constant is not readily used in calculations because the two proportions are different, mass and volume. If, however, one takes the mass of water vapor and compares it to the mass of the dry air of the oxygen and nitrogen molecules and included within that totality is the mass of water vapor as well, then an accurate analysis may be reached, termed *specific humidity.* Taking the mass of water vapor and comparing it to only the masses of oxygen and nitrogen, a mixing ratio will be established (Ahrens, 1994:115). Interestingly enough, as long as the specific humidity and mixing ratio is based upon no additional intrusions or the removal of water vapor, then they will remain constant and in equilibrium as a mass relationship, regardless of parcel volumetric contraction or expansion.

One of the parameters that vary due to vertical dynamic motion is temperature. As the volume within the parcel expands, there is an exchange of heat between the parcel and the surrounding air because of molecular motivation pushing against the parcel boundary. To accomplish this outward push, the parcel molecules require some of their own energy to accomplish expansion, and there will be less motion between them. A consequence of this lower molecular motion in turn causes the internal heat that would otherwise be generated from compression to lessen, cooling the internal atmosphere of the parcel. It is this difference between the internal air parcel's temperature and the surrounding air that provides the dynamic of whether the air parcel will continue to rise, remain at that altitude, or fall, and for this reason air temperature at various altitudes is extremely important in any analysis concerned with atmospheric mixing. At a specific altitude, should two similar-sized parcels be examined that are at different temperatures, the colder the parcel temperature, the more molecules it will have within it, thus the denser and heavier it will be in contrast to a parcel with a warmer temperature. Because the parcel's temperature-to-density relationship causes the temperature to fall because it is colder than the rising air surrounding it, a condition of resistance to upward mobility is in effect. This condition is termed *stable air.* An unstable condition would therefore exist that would permit a parcel's upward mobility because the parcel's internal density is less (warmer temperature) than the surrounding air.

A relationship now can be analyzed that encompasses the variation between surrounding air and a parcel's. As altitude is gained, the atmosphere will cool; this lowering in temperature with altitude is called the environmental lapse rate and is a predictable function. The parcel's temperature changes, too, and that condition applied against the rate of temperature change in altitude can yield information pertinent to total atmospheric behaviors.

Bear in mind here that although the air parcel may be at the same temperature as the surrounding air at altitude, the pressures within the air parcel and the atmospheric pressure exerted by its atmospheric environment may not be equal (Seinfeld and Pandis, 1998:767). The process that covers this morphic change of an air parcel as it cools through expansion or becomes warm due to compression is termed *adiabatic* and will remain at a constant rate of ten degrees Celsius for every 1,000 meters in elevation change only if the air in the parcel is unsaturated (less than 100 percent humidity), thus the terminology *dry adiabatic.*

Should the air parcel have water vapor within it, then as the parcel falls and becomes internally warmer through compression, the water vapor will evaporate. Because the evaporation of water vapor causes a cooling, if the rate of cooling offsets the heat generated by the parcel's compression, then the moist adiabatic rate is achieved, which describes the thermodynamic process of entropy (Emanuel, 1994:131).

This study has focused upon the micro or single parcel of dry and moist air, and until now any analysis of multiple molecules has not yet been addressed. For our consideration, two areas over two geoterrestrial locations can be modeled to investigate changes of pressure, density, and temperature on a large scale.

Atmospheric Relationships

An area will be examined over two surface locations and will be in columnar configuration, enabling density investigations to be relevant with altitude. Each location, although distant from the other, will be at the same surface altitude and have equal surface pressure delineating each column at both locations to contain the same number of molecules (mass), also that the distance between molecules is such that the pressure is equal and remains constant throughout columnar heights.

If the temperature at the top of one of the columns becomes colder than the top of the other column, an area density imbalance occurs. The column containing colder air will remove molecular energy, causing those molecules of air to fall; this activity establishes a denser region at the surface, which will increase that pressure. The warming of the second column will increase its molecules' motion by imparting energy to them, causing that containment to expand upward in columnar height; this results in fewer molecules at the surface, lowering that pressure.

These situations lead to a fact in meteorological sciences, that is, in order for two surface pressures to be equal at different air temperatures, the altitudes of the air columns must vary. A colder air column will be less in height or altitude due to its descending molecules to a state of compactness than would a column of warmer air that has expanded because of molecular motion, even though the number of molecules has not changed. Because of this physical fact, it takes a shorter column of colder air to exert the same surface pressure as a warmer column.

Also note that as you increase in altitude within the colder column, the threshold level of molecular density will be less sooner and cause a rapid decrease in atmospheric pressure. On the other hand, the warmer column of air, being greater in height, will sustain a small variance in pressure during an ascent, because the pressure of internal molecules are not clustered closer to the surface but are farther apart and spatially equal throughout.

Due to the fact that pressure is an indicator of molecular density and since pressure is based upon a downward effect and representative of the molecular densities above a reference point, approaching the top limitation of the cold air column, there would be fewer molecules between the sample point and the top cold threshold barrier of that column. This reference point would experience less pressure than at that same point in height (altitude) within the warmer column if air. The increase-in-pressure observations between the two columns of air when taken at the same altitude within the warmer column are due to a greater remaining molecular density between the sampling point and the warm column's top threshold point of air. It is because of this temperature-to-density-and-pressure relationship that a high atmospheric pressure will be associated with warmer air and a lower atmospheric pressure will be associated with colder atmospheric air at altitude. In addition, the colder molecular air mass density will not be as high in altitude as a warmer molecular air mass. In the verification of two horizontal differences in temperature creating a horizontal difference in pressure at altitude, molecular motion created by that fact can now be analyzed.

With reference to figure 6.9, and the preceding statements denoting equal horizontal altitudes with respect to pressure and temperature differences, if the systems (each column having the necessary parameters met for pressure, temperature, and altitude) are brought into proximity to each other, an additional dynamic is created. That dynamic is observed as a molecular velocity from one system to the other.

Interestingly, the surface of our planet is not uniform in topographical heights. Many mountainous places that are near the ocean have heights of several thousands of feet, and from a planetary scale, these variations in height can be extremely diverse. As studied, the higher in altitude one travels, the lower the temperature expectations, due to the lapse rate (fig. 6.10). Also, from a planetary scale view, there are regions on the planet that will always vary from other places on the Earth, and those variations are always warmer with respect to other proximities (tropics versus polar). The warmer regions will have molecules of the upper atmospheric layer pushed vertically by the expansion upward of molecules nearer the surface, where the warmer temperature instability originates. As the displaced warmer air expands upward, it will eventually reach a maximum height, come under the effects of latent heat loss, lose buoyancy, and fall back to earth.

The displacement of molecules leaving the surface to convect upward because of warming leaves an absence from that mass, and this imbalance will cause the region of density to be infused by the molecules that have descended back to the surface. The point at which the cooler molecules have fallen earthward will begin to increase that regional pressure to a point that surpasses the pressure at the center of thermal convection. With the pressure from additional molecules now higher than that of the lower mass density point of convective origin, an atmospheric flow toward the

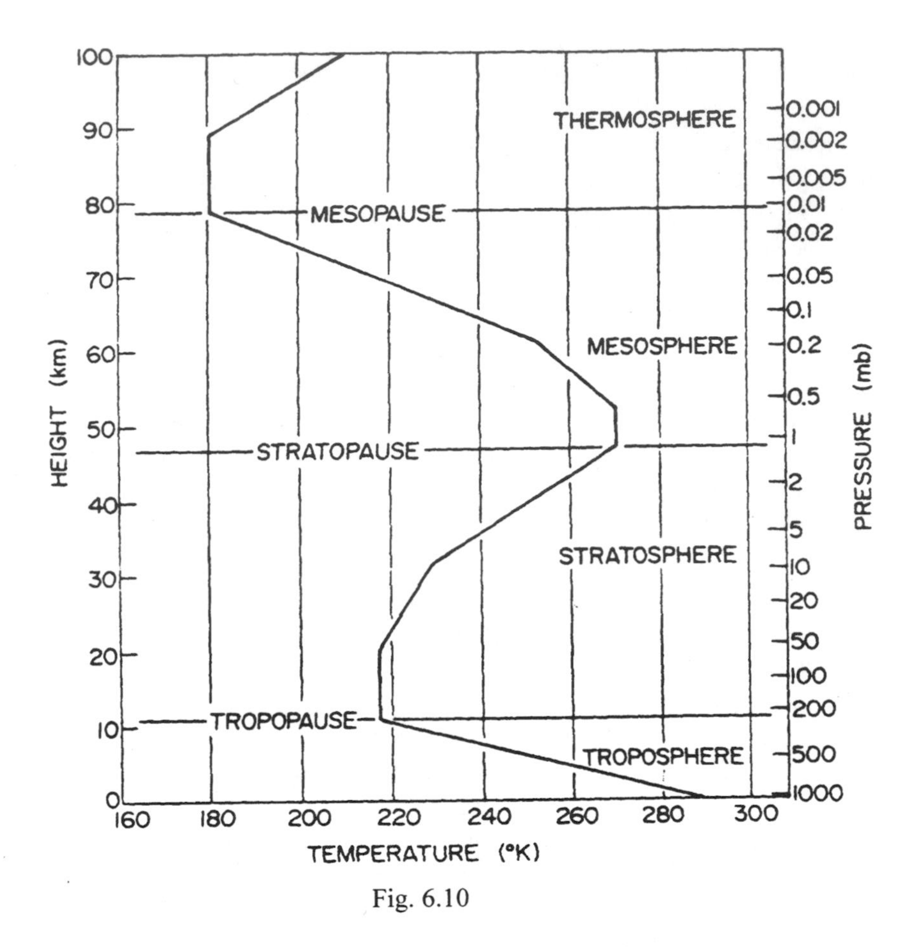

Fig. 6.10

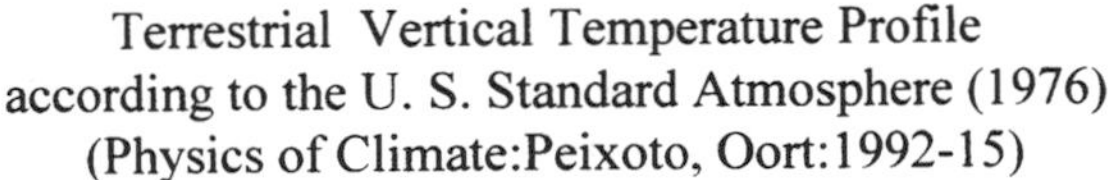

Terrestrial Vertical Temperature Profile
according to the U. S. Standard Atmosphere (1976)
(Physics of Climate:Peixoto, Oort:1992-15)

region of lower pressure is initiated. That horizontal molecular flow will correspondingly have a velocity that is determined by the distance and pressure magnitude between the two pressures. As an example, two pressure differences that are one half-pound per square inch in variance will, at 500 miles apart, accelerate the atmosphere to 80 miles per hour in three hours. If the distance between the two differences is doubled to 1,000 miles, the wind velocity will be one-half, or 40 miles per hour, after three

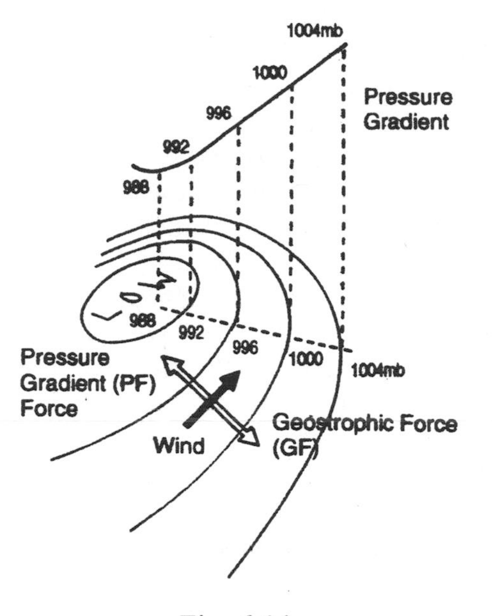

Fig. 6.11

Pressure Gradient, Strophic's and Wind Direction
(The Weather Handbook:1999-100)

hours (Williams, 1997:35). Although this example is an approximation and does not consider surface friction and planetary rotation, it will demonstrate how the horizontal force structure operates between points of variable pressure systems wherein, with the air parcel as a whole being in motion, its change in position is defined as velocity (Dutton, 1995:175).

Because a variation of pressures can and does exist over micro- and macroregions, studies with respect to their variances play an important role in any analysis of meteorological systems. The fact that a variance exists at all is indicative of two different values and a magnitude relationship between them. Variations by degree, either upward or downward, are termed *gradients.*

Since pressure is a variable, it can be expressed as a gradient if one assessment point with respect to the other is determined by graduations. Those graduations will comprise the gradient over a specified distance, thus termed a *pressure gradient.* These assessments are identified in millibars, referenced earlier, and are illustrated in fig. 6.11. An establishment of parcel motion, i.e., wind, which is moving through an accelerative condition, attaining a velocity that is relative to the differences in atmospheric pressures, assessed in locations separated by some distance, can therefore be applied to the terrestrial surface picture as a whole. And one of the major influences that will govern the parcel in motion having a velocity due to pressure differences is the rotation of the Earth.

Differences in equal pressure are not identified the same as points of assessment where the atmospheric pressures are equal. In cases where equal pressure measurements have been taken over the surface over some distance, they can be joined by a connecting line called an isobar. Isobars would be parallel if the Earth's rotation was not factored in, and the parallel separation between the connecting lines referencing points of equal pressure is then relative to the gradient between their values. If the parallel lines between two equal pressure measurements are close together, than the pressure gradient would be greater than if the connecting isobars were separated in parallel by being farther apart. The parcel of atmospheric motion eventually finds a balance between two of the isobars in its velocity and direction. Since the parcel has realigned itself to an equality of forces upon it and the rotation of the Earth's surface friction has not yet been employed to the parcel as a force, where only the pressure gradient is being considered, the motion generated in this manor is termed the *gradient wind.*

However, the Earth rotates, and because of this strophic (twisting) influence termed *Coriolis force,* the parcel in motion is affected and will manifest that influence as it progresses between the gradient isobars. It is here that the Coriolis force will have a horizontal deflective influence and modify the parcel's motion dynamic (parcel's forward vector) (fig. 6.11). Because the parcel will be under gradient influences, the gradient isobars themselves will effectively control some of the parcel forward adjustment. However, as a consequence of the Coriolis force the isobars will curve around the centers of pressure, allowing the direction of the parcel to continually adjust to the curvature influence, always at a right angle to the parcel's motion. Once the parcel has completed its foreword vector alignment with respect to the strophic influence and pressure gradient equalization, then the motion becomes a geostrophic wind. The deflective influence of the force will be to the left (up) from west to east across the Northern Hemisphere and right (down) in the Southern Hemisphere (1998:105).

The Coriolis force is strongest from the surface to about one kilometer or above in height, due to the absence of the terrestrial surface friction influencing the gradient pressure isobars into a curve. At altitudes above one kilometer, the minimal strophic influence allows the parcel to move as a gradient wind between isobars that will be reflected in a more parallel and straighter line configuration. In addition, the geostrophic wind is inversely proportional to latitude, and because of this consideration the strophic component of the vector system upon the parcel will approach zero with proximity to equatorial latitudes. For this reason and because the vector will be minimal, the wind that would be experienced at the surface level would be more in line with a gradient force structure (Chorley, 1998:106).

In consideration of isobaric equalities representing horizontal variable pressure gradients, where the gradient differences do not differ between two lines representing pressure equalities we would expect to see the spacing between them in parallel fashion; if, however, the gradient difference increases at some point between them, relative to the isobars, then the gradient spacing would indicate the increased difference by moving closer together. The two parallel lines of equal pressure would bow into each other; this narrowing of isobars is termed *confluence* and is demonstrated in the atmosphere by an increase in air velocity. If the isobars in a parallel configuration are altered by a decrease in pressure difference at some point along them, then the isobar lines will bow outward and become farther apart. This spreading configuration is termed *diffluence* and atmospheric behavior would be observed as a decrease in air velocity (Chorley, 1998:108).

When atmospheric parcels move over a terrestrial surface, the parcel will experience frictional differences, illustrated through a velocity change. If the parcel encounters land after being at some velocity over water, the friction of the land will cause a deceleration to occur; correspondingly, if the parcel moves from land to a water surface, the lack of friction from this transition will cause an acceleration of that parcel.

The compressibility of the air as it slows its velocity from transitions of water to land is called convergence, and if the parcel moves from land to water, a divergence of air is in effect, due to expansion from the acceleration dynamic. When discussions of motion encompassing both divergence or convergence of air are entertained, the fundamental cause for their generation is based upon two pressure systems that are in proximity to each other replacing their horizontal air mass losses through the compensation mechanism of convecting or descending of vertical motion (Gordon et al., 1998:110).

In the cases of a low-pressure area and a high-pressure area next to each other, both convergence and divergence are in effect. Convergence of air into the center of the low pressure at the surface is a function of replacement of the divergent air leaving the top after vertical ascension. The divergent air moving outward at altitude from the low pressure area at the top moves to a convergent definition at the top of the high-pressure area, replacing the divergent air leaving the bottom at the terrestrial surface of the high pressure column after its cooling and descent.

7
Protection from Above

Radiative Atmosphere

Discussions have been focused upon distinct areas of low- and high-pressure cells convecting molecules in vertical and horizontal motions. Although the dynamics of divergence and convergence have been discussed as physical motions, a factor that has up until now been absent is the relationship that water vapor and the dry air components of nitrogen and oxygen have with thermal influences. Patterns of lows and highs must in some way reflect that experience of stimuli from this additional atmospheric bond reflex and photonic capability. As studied, the chemical and atomic bond relationships in concert with diatomic possibilities by polarization to a magnetic field must be of some significance to any atmospheric structures around us. In addition, there must be some consequence because of the atomic exposure to electromagnetic energy that is observable without any special instrumentation.

The best observations of the possibility that our atmospheric constituents react to electromagnetic energy can be made from transitions of a clear to cloudy day. It is obvious that some energy transfer is taking place due to the formation of clouds alone and that they create a shadowing effect from the sunlight. This observation is indicative that water vapor has some absorption quality and is therefore reactive on an atomic level. In Chapter 2, the identification of electrons and their orbital frequencies was discussed with a relationship to the emission of electromagnetic wavelengths that identify a particular atomic configuration. It would seem in this discussion that the best place to start and expand that knowledge of reactions would be to apply it to the atomic molecule of water—that is, after all, what clouds are—and eventually the remaining atmospheric compositions and their reactions to the sun.

The sun we see every day offers radiant heat energy, received by the Earth some 93 million miles away and eight minuets later. The sun contains 99.9 percent of the mass of our entire solar system and is so large that it could contain all its planets within it. The sun has a composition of 92 percent hydrogen, 7.8 percent helium with less than 1 percent of oxygen, carbon, nitrogen, and neon (1995:36). The energy we receive on the Earth is released outward from the sun as ionized gas called solar wind, traveling at 435 miles per hour with a density of between 10 and 100 particles per cubic centimeter (1995:46). In this instance, because the sun is producing wavelength energies to such an extent that they propagate outward through space to illuminate this planet, they are said to be radiant. In our situation, the received portion that eventually comes into contact with the troposphere is considered under the general term of *radiation.*

The radiation in wavelengths from the sun extends from the very high frequencies in cycles per second or as extremely short wavelengths of gamma and X-ray and ultraviolet to the longer wavelengths of visible, radio, infrared, and microwave. The total bombardment of electromagnetic wavelength with respect to the quantum of associated energy can be seen in figure 7.0 as the average irradiated distribution at the top of our atmosphere and at sea level. The shaded area is of concern to this study, as it represents wavelength absorption references for atmospheric components at the sun's zenith. Chapter 2 discussed the principles behind electromagnetic theory as it could be applied to atomic configurations. Here the application of electromagnetic theory is not as a definition of orbital configurations but as an effect on the structures of atmosphere containment. As a living species, we contain a mass that is made of matter through the formation and adhesion of atomic configurations. Because atoms will be affected by their proximity to electromagnetic fields and wavelengths of specific angstroms, that effect is most important when the sun's bombardment contains spectral properties that will kill.

From an atmospheric standpoint, we are surrounded by radiation of some magnitude not only from the sun's bombardment but also from energy that is radiated back into space from this planet. Because there is a form of stabilized balance between the radiative properties of the sun and Earth, that radiation being received by the Earth must also have balanced against it as an equal amount outgoing energy from the Earth back into space. It is important to remember in this application that the higher a body's temperature measured in Kelvin, the higher the frequency and shorter the wavelength in angstroms that will be emitted (Planck's Law).

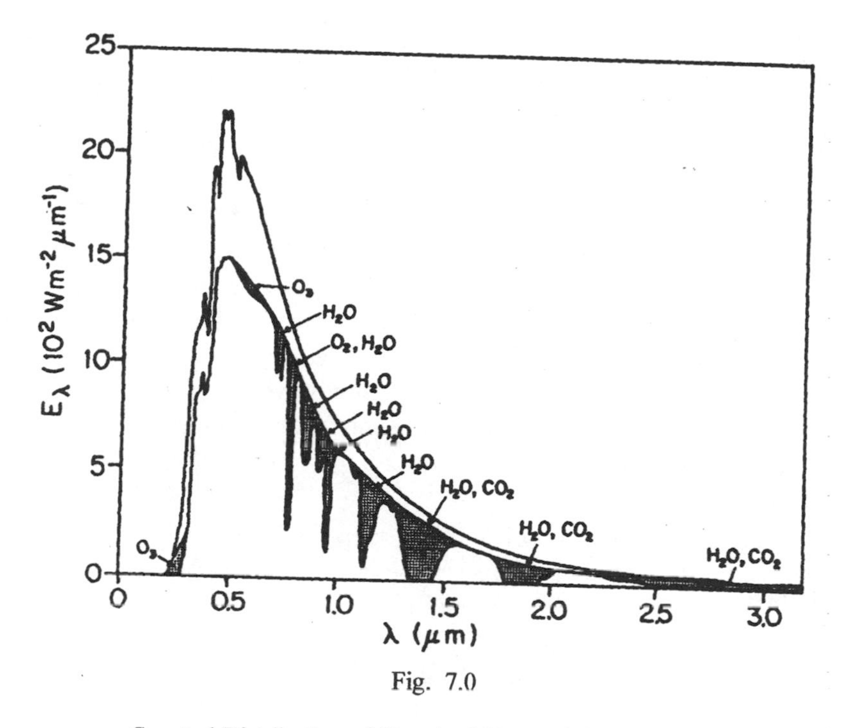

Fig. 7.0

Spectral Distribution of Received Terrestrial Solar Irradiation
(At Altitude and Sea level)
Permission: United States Air Force

Of course, the sun's temperature is vastly hotter than the Earth's surface; therefore, the irradiation spectrum in both energy and wavelength will demonstrate this fact. In figure 7.0, this can be illustrated by the significant increase in energy (left side, vertical) and the analogous shortness in wavelength. The outgoing radiation that is generated by the Earth's surface is primarily in the infrared region of wavelengths less than ten micrometers (um) The solar interests that would affect our climactic systems are in the regions of .1 um to 2.0 um, representing ultraviolet, infrared, and of course visible light. Interestingly, because the sun is at some distance from the Earth it can be viewed as a point of source and because of that the light bombardment is almost in a parallel format. In contrast to that radiation, the Earth is so close and large to us that the radiation that is emitted from it will appear to come from all directions at once due to molecular densities all reacting as extremely small suns radiating thermal diffuse energy (1992:92).

In considerations of incoming solar radiation as 100 units, 16 percent (units) are absorbed by the atmospheric constituents of water vapor and ozone, 4 percent (units) are absorbed by clouds, and 50 percent (units) are absorbed by the Earth's surface. The atmosphere will backscatter 6 percent (units), 20 percent (units) are reflected back to space from clouds, and 4 percent (units) will be reflected from the Earth's surface. The last 30 percent (units) that the atmosphere and Earth reflect back are not considered active stimuli within the climate system. The balance that the Earth radiates back into space come from 20 percent (units) of longwave radiation emitted from the surface, and of that, 14 percent (units) are absorbed from water vapor and carbon dioxide within the atmosphere, and 6 percent (units) make it to space. Of infrared radiation that is emitted from the atmosphere, 38 percent (units) are from water vapor and 26 percent (units) are from clouds (1992:94).

As far as the spectrum of bombardment is concerned, 99 percent of solar radiation that reaches the Earth is in the wavelength region of between .15 and 4.0 um. Nine percent is defined as ultraviolet (less than .4 um), 49 percent is visible light (.4 to .8 um), and 42 percent will be in infrared (1992:98). Because of the Earth's tilt on its axis, the solar radiation that bombards the equatorial regions will differ from the radiation that reaches the surface at higher latitudes. Photonic particles travel farther at the higher latitudes, so the absorption probabilities are higher for certain wavelengths than for equatorial regions that receive the energy in a more vertical manor. The higher latitudes provide for a longer path through the atmosphere, which promotes molecular absorbitivity.

It is this absorbitivity that is important to atmospheric understanding, because within that understanding will come the application of atmospheric bond reactions that sustain us as a lifeform.

Photodissociation Shield

There is something that atoms do when they come into contact with a wavelength that affects the bond between them. In Chapter 4, the discussion was of how the molecule would vibrate, specifically the rotation or vibration of the atomic structure depending upon bond characteristics and at certain levels of absorption or emission. We are interested in only an apportionment of solar irradiative absorption for the moment and a distinct effect upon the oxygen molecule and atoms. The first part of this investigation is the consideration of that part of the radiation spectrum that would be considered harmful to us as a species and how the atmosphere provides for our protection from it.

One of the reasons that any consideration of radiation is done is because of the effect certain wavelengths have in their ability to alter matter, meaning it is the chemical bonds that are affected to such an extent that the molecule is broken into its individual atoms. Such a case is seen between the oxygen molecule and the part of the solar spectrum identified in the ultraviolet wavelength. When ultraviolet radiation reaches the upper atmosphere, that is, the stratosphere that extends to fifty kilometers in height, because of this height, the molecules of atmospheric gas are much fewer than would be found at lower altitudes due to gravity. Even so, the oxygen molecule does exist in sufficient abundance to be affected by this particular wavelength of solar energy. There has been a great deal of focus on the photochemical reaction between these two subjects because of the health risks associated with human ailments such as skin cancer, cataracts, and other atomic dissociations that wavelengths of this nature will produce.

When the oxygen molecule (O_2) is irradiated by this wavelength, it will break apart into monatomic oxygen, and the individual oxygen atoms will then recombine with existing oxygen molecules to produce the ozone molecule, or O_3. Ozone is formed naturally at this altitude because of this process, and this is called the Chapman Cycle (fig. 7.1) (1998:4). It is precisely because the ozone molecules will be dissociated into monatomic and molecular oxygen through the absorption of ultraviolet wavelengths of solar energy that is important and therefore prevents the harmful radiation of the ultraviolet from reaching farther down into tropospheric levels that protect biological life. Ozone production, however, does have a downside; that is, ozone is toxic to humans and is of consequence because of the pollution factor, which will be addressed later. The oxidants of O_2 and O_3 have very high bond energies and therefore will not be reactive, as would molecular structures that do not have this characteristic (Jacob, 1999:200). They are, however, reactive to H-containing molecules. In consideration of UV reactions within O_3, it is necessary to describe for clarity the process of ionization, for it is that process that creates the necessary oxygen components that sustain the Chapman Cycle, and to determine what happens within the atmosphere as a result.

In this understanding, before any activity of the atomic molecule can be initiated, the first part of the process is the capturing of the wavelength

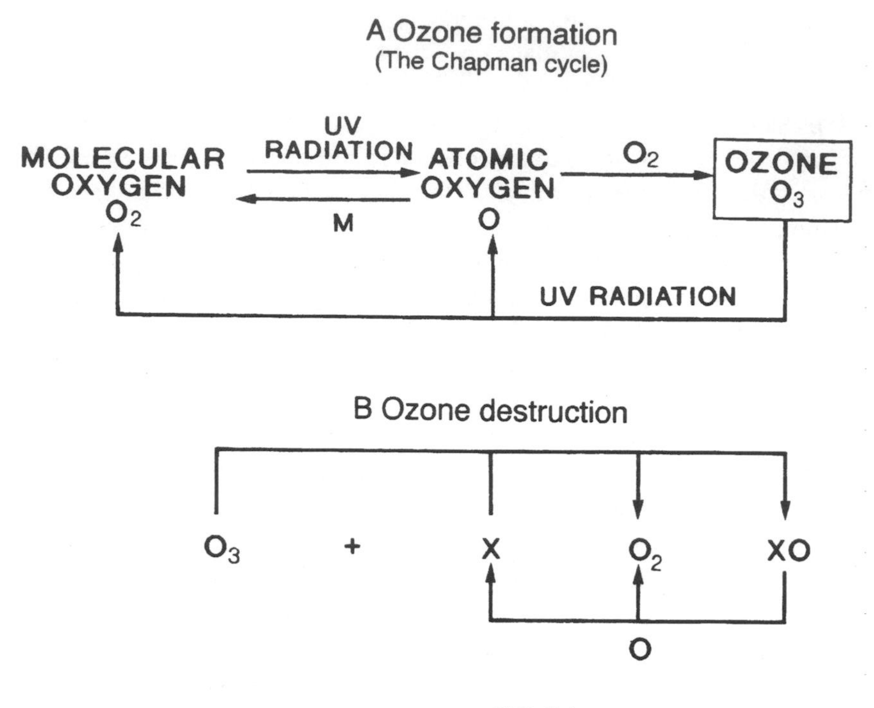

FIG. 7.1

The Chapman Cycle of Ozone Formation / Destruction
(Atmospheric Weather & Climate:1998-5)

particle. When an atom is in proximity to a particle with an energy level that is reactive to an electron in orbit, molecular dissociation occurs when the energy of an incoming photon is greater than the binding energy that holds the molecule together through its electron field and when a resolve to the inner atomic system ejects its components creating a net charge of either positive or negative quanta (1996:398; Seinfeld and Pandis, 1998:141). In the ozone molecule, the O to O_2 bond energies are roughly 105^{-1} kilojoules per mole and can be identified with the lowest energies that will cause photodissociation and are below the visible spectrum in wavelength. The absorption of wavelengths here is important to our case only if the upper and lower levels of energy within the molecule are separated by an amount that is equal to the energy of the photon. The O_2 molecule will absorb wavelengths that are less than 200 nanometers, and

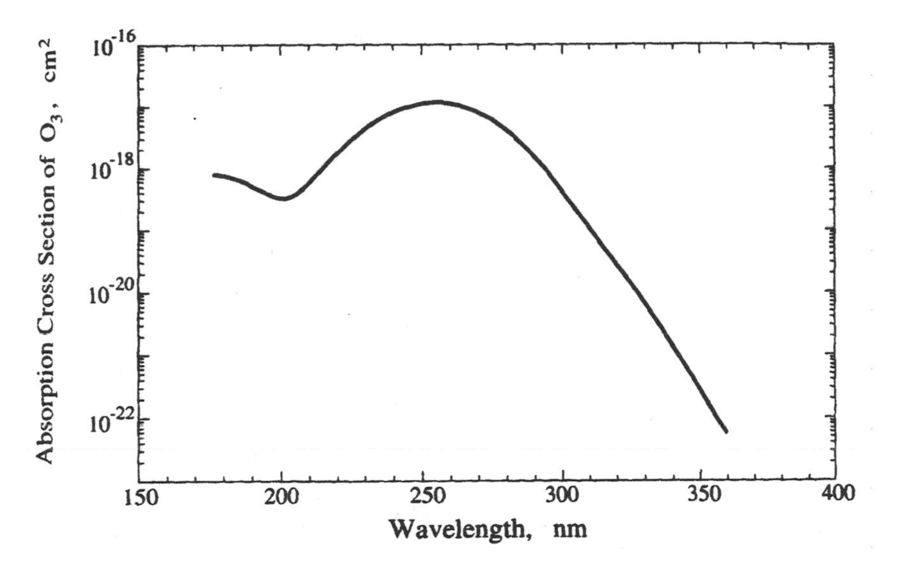

FIG. 7.2

Wavelength Absorption by the Ozone Molecule
(Atmospheric Chemistry and Physics:1998-147)

because of this the farther into the atmosphere shorter wavelengths travel, the more of that wavelength will be absorbed and removed. As a consequence, the wavelengths of light that reach the troposphere after stratospheric absorption are limited to being longer than 290 nanometers (Seinfeld and Pandis, 1998:146). This photodissociation and photolysis can be seen in figure 7.2, illustrating the greatest absorption for the O_3 molecule at about 250 nanometers in wavelength. If another photon of energy equal to the first irradiation arrives before that electron has fallen back from its excited state, then that additional absorption may cause the electron to be removed from the system and ionization will occur. Collisions, too, will cause electron ejection if the imparted level of energy is greater than the First Ionization Potential, which is defined as energy in eV required to remove the least bound electron and, as a consequence, disassociate the structure. In cases where additional subshells exist beyond one, then higher thresholds of energy would be needed (1991:561).

It is important to know that not all photon absorption will initialize ionization. In situations where low binding energies exist, then the percent

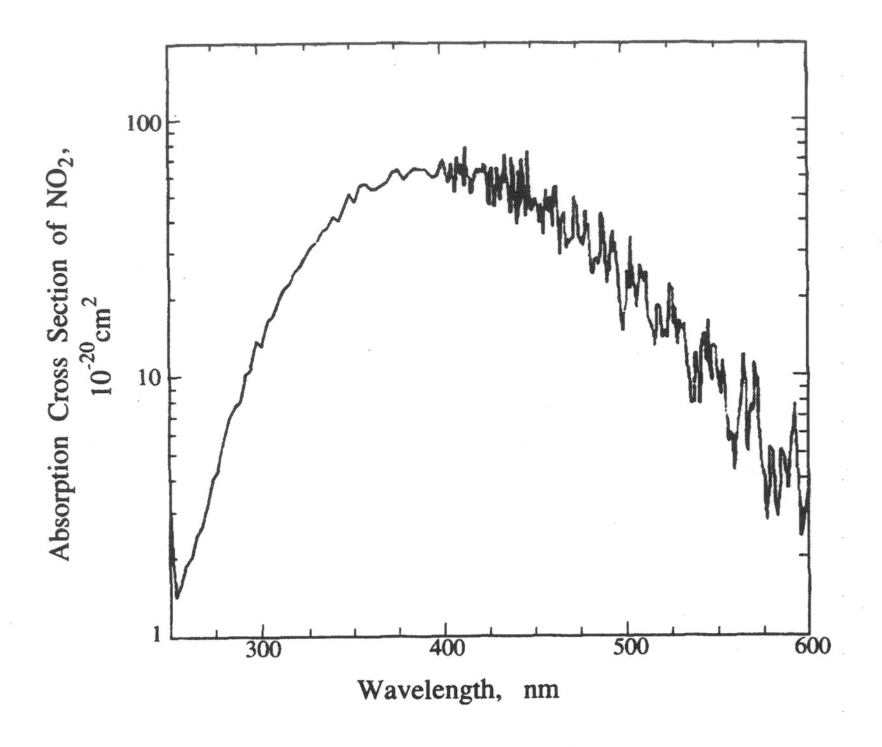

Fig. 7.3
Absorption Spectrum for the cross section of NO_2 at 298 K
(Atmospheric Chemistry and Physics:1998-148)

of disassociation will be high. Also, the level of ionization is very closely related to the cross section of the atom under irradiation bombardment.

Nitrogen Photolysis

Seventy-eight percent of the atmosphere is nitrogen; 21 percent of the atmosphere is oxygen. There is something significant about these two in combination. Not all the oxygen is from the dissociation of ozone; the NO_2, or nitrogen dioxide molecule is just as important in this process as well.

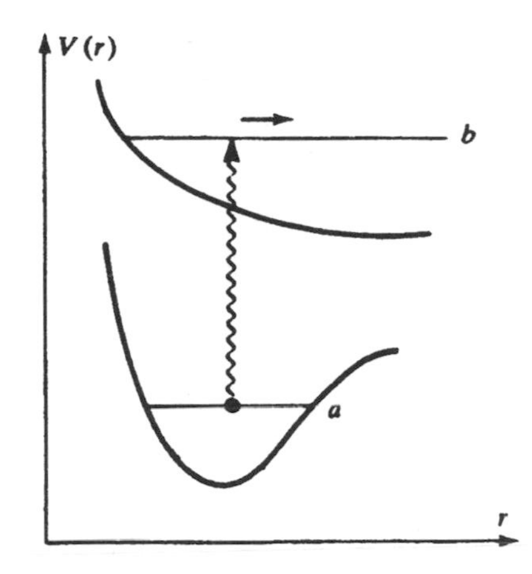

FIG. 7.4
Recombination / Dissociation Depth Energy
(Atom - Photon Interactions:1992-82)

Nitrogen dioxide will produce an oxygen atom at the expense of dissociation within the nanometer range of between 280 and 400. Also, it is very valuable in the absorption of ultraviolet and visible wavelengths (fig. 7.3) between 300 and 370 nanometers at 298 Kelvin. Since the bond energy between oxygen and nitrogen is about three hundred kilojoules per molecule to the minus 1 power, its corresponding wavelength for dissociation to occur is about four hundred nanometers. Because the ground state of the nitrogen dioxide molecule varies and is not precisely stable, that variable is reflected in figure 7.3 and illustrated by an increase in perturbation prior to the graph's decline.

In addition, since a change in the chemical bond relationship is necessary to establish the molecules' dissociation, then that chemical alteration is considered to be an energy translation. The total energy then includes the thermal and can be associated with nuclei motion contained within the internuclear distances concerning such molecules (Cotrell, 1954:5). In consideration of this energy, as the atoms decrease the distance between them, their respective nuclei-electron system energy decreases and is considered to be the minimum potential energy for the nuclei; as the proximity decreases farther, that energy level will increase at a rapid rate (fig 7.4), and as is illustrated, the depth to which the energy decreases is the characteristic energy that defines the molecule when the atoms combine. With reference to figure 7.4, the vertical represents the energy associated with nucleon particles, whereby in a dissociated configuration the horizontal represents the nucleon distances and in the case of absorption the excited state is (b) and the ground state is (a).

The Magnetic Component

In Chapter 2, some discussion was entertained regarding electromagnetic properties that influence the dipole structure and, to some extent control particle charge behaviors. This flow of electrons is more important than one would normally expect since (1) the field is almost invisible to us optically and (2) an influence from the field about the Earth has properties that atmospherically would have any meaning to anyone beyond a rainy or clear day. I say this because of solar bombardments regarding extremely high-particle energies and because few in the populace would even consider an Earth field such as this in existent at all. At this point in the study, I refer to figure 2.24. The flow of particles can be seen in this representation as a field surrounding a dipole structure. This is very similar to the actual field that surrounds us and consists of the invisible atmospheric component that is in many respects more important than ozone is to photolysis. This field, within and a distinct part of the atmospheric entirety, which is so important, must have an origin and consist of matter if it is to exist at all. Having made that statement, I can further affirm that this atmospheric component in question does not consist of molecules but energy and not in the form of an atomic structure but in the form of nucleon quantums.

The first step in the exposition of the field about us is to define the process by which it is created, then to apply that creation of quantum energy to the Earth as an atmospheric component. We have previously defined the atmosphere as a molecular density that has the ability, through constituent matter, to remove ionizing radiation within stratospheric and tropospheric boundaries by the process of photolysis. The essence of this thought is to analyze the importance of that irradiate removal by its association with our survival as a lifeform. The molecular exchange between certain wavelengths of particle energy and its reactiveness with molecular existences then define its importance to us. Further, the importance to the existence of matter in that it will disassociate when in contact with such high-energy particles as the ultraviolet would seem to give impetus to further thinking about such matters when higher energies exist far beyond that wavelength. It is precisely because higher irradiate bombardments occur that further analysis of a molecule's inability to remove that higher spectrum must be given consideration. Higher orders of wavelength are in reality, and because the tropospheric and stratospheric molecules only absorb specific wavelengths that do not include that reality, there must be something else that assists in our protection, if protection is indeed the reason for an existence beyond a chemical reaction at a molecular level. This existence is the subject that we seek to expose, how is it produced, and, most important, why. By what process does the existence depend and, how do we know that the reality of what we seek is beneficial?

The expositional sequence with concern to the presence of atmospheric magnetism, would be to analyze the Earth structure for evidence establishing any formations that are necessary and applicable to dipolar configurations. That is, to determine if the Earth itself consists of a concentration of either predominantly protonic or electronegative quanta. This is of paramount importance, for it would identify the fact that two quantas separated by a finite distance will create a charge potential and it is that potential between quanta differences that establishes the nucleon influence through a dynamic proximity to and from it.

Terrestrially, the Earth is not totally solid (fig. 7.5); it has a surface (litheosphere) that represents for the most part a structure that has cooled from a liquid or molten material (magma) originating from deep within. The pressure at the lower depths of the Earth is quite great; a consequence of this pressure upon the material subject to it is heat. In addition, that temperature at great depth is exacerbated by the radioactive isotopes of uranium 235 and 238, thorium 232, and potassium (1982:6). Intense heat at these levels is not enough to warrant the explanation of quanta polarities; however, high thermal temperatures coupled with the fact that there is both heating from below and a cooling from above within the presence of a gravitational field will induce the material to become gravitationally unstable. This instability produces a circumstance whereby the colder surface magma descends and the hotter magma ascends, constituting the engine for geodynamic motion, and although magma dynamics can be factually determined and important enough, this is still insufficient to develop quanta polarities. The molten constituents themselves must therefore be a factor in this analysis. The generation of a force that will influence nucleonic motion in this particular situation must be of a certain formation of

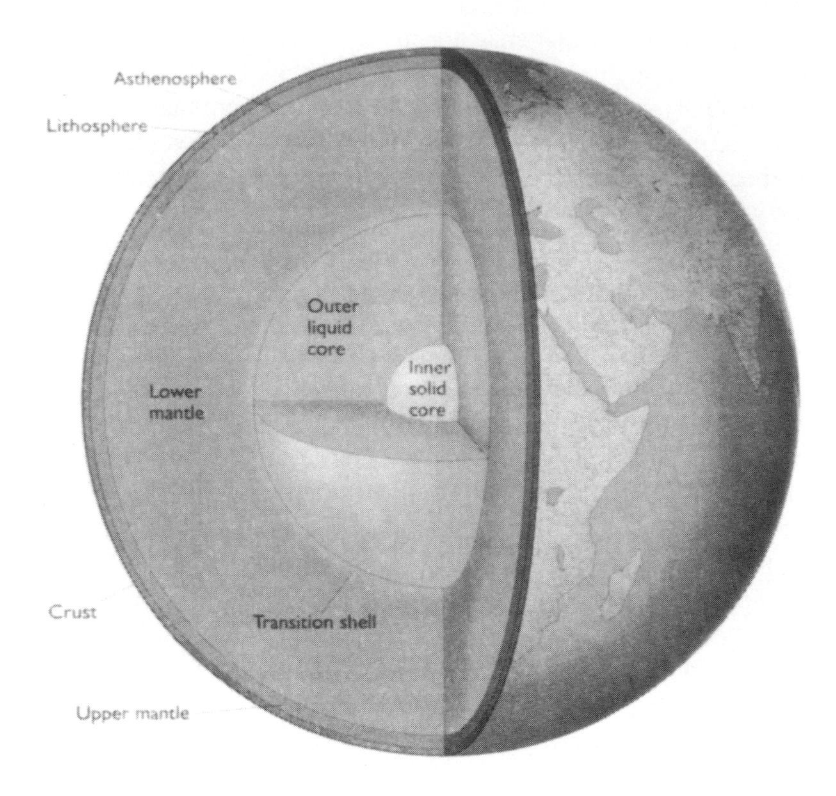

Fig. 7.5
Cross Section of Earth's Gross Structure
(*Earthquakes and Geological Discovery*: 1993-127)

matter that contains properties that would meet the atomic requirement for a signature to an electromagnetic characteristic. The magma in solid form and the magma in molten form must therefore have between them those properties in a dynamic that can be considered the primary focus.

The fact that magma itself can be in two states is intriguing and presents a question of the morphic conditions having some influence on the state changes themselves insofar as an electromagnetic property is analyzed. The term *magma* means "melted rock" (1983:9) and at some point in depth, pressure, and temperature the changes between the states of solid and molten occurs. As the heated magma rises, and although its temperature remains nearly constant, the pressure around the magma area lessens and decreases. This in turn affects the solidus temperature, or the point in temperature at which the rock starts to melt. As a consequence, the higher ascending magma rises, the lower the solidus temperature becomes. This heating and cooling depth relationship takes on significance when applied to the magma constituents of elements within. Certain temperatures, when applied to magma from a heating to cooling phase can determine if any magnetic properties within the rocks' formation process at this point will be preserved. If magma minerals are heated above a specific temperature, called the Curie Temperature, then any possibility of electromagnetic retention will be destroyed, because the electron distances will have been increased to a disassociative state and preclude electron sharing. If, on the other hand, magma is cooled to the formation of rock below the Curie Temperature in the presence of a magnetic influence and if the rock constituents are such that a magnetic characteristic can be imprinted, than the rock will posses thermoremanant magnetism (1983:21). The rocks in which we are interested would be required to contain some ferrimagnetic matter, that is, matter that on an atomic level will have their polarities at each dipole end parallel to each other; this unique state will allow them to become aligned in the presence of a magnetic field line of force most easily and quickly. This is important when the rock cools and the orientation of the dipole is expected to remain in that polar alignment when outside of and no longer in proximity to the force lines' influence (1995:150 [fig. 7.6]), because all substances at an atomic level are dipoles due to the spin and orbital paths of its electrons along with the electron count. It's this electron count that gives the susceptibility to realignment. The unpaired electrons in the outer shell of the atom constitute an unstable situation, and this is considered a paramagnetic condition. When the atom with its unfilled and therefore incomplete shell configuration comes into contact with a magnetic line of force, the unpaired electrons will advance a spin, resulting in an equally represented magnetic field (electrons in motion), and remain in that spin configuration forming a weak field on their own when removed from the influence (1995:150). The substances that are part of the magma constituent and will remain in a polar alignment after field subjugation are called iron, cobalt, and nickel.

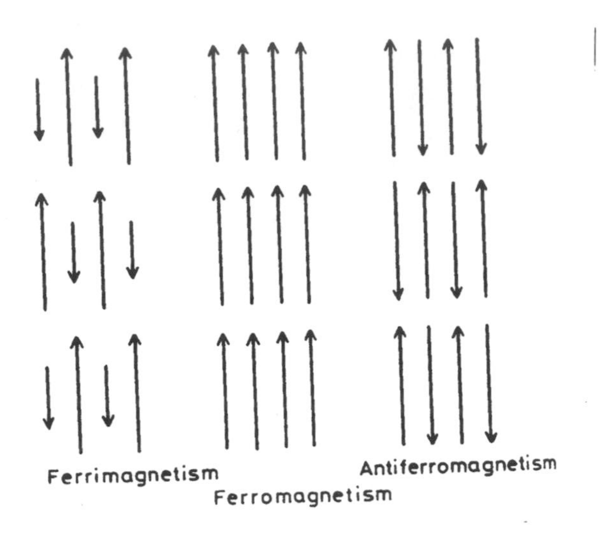

Fig.7.6
Elementery Dipole Alignment
(Introduction to Geophysical Exploration:1995-150)

The Field

With this information, a representation of convective motion by molten rock can be placed in the context of a magnetic dynamic. Because the temperatures at the Earth's core are higher than any Curie Temperature for known magnetic materials, the reasonable solution for any field property must come from the atomic coupling of paramagnetic minerals trapped within the molten rock substance by an interaction with convective motion and the core. The center of Earth's spherical form is solid from the structure of an iron core in existence, which also provides the electromagnetic impetus.

It is therefore this interaction of the Earth's solid iron core and the outer molten fluid that would be a most reasonable assumption for the influences' origin, and the result can be seen as the approximation in figure 7.7. An electromagnetic dipole configuration representing the Earth's geomagnetic field lines of force is in agreement with approximation equalities

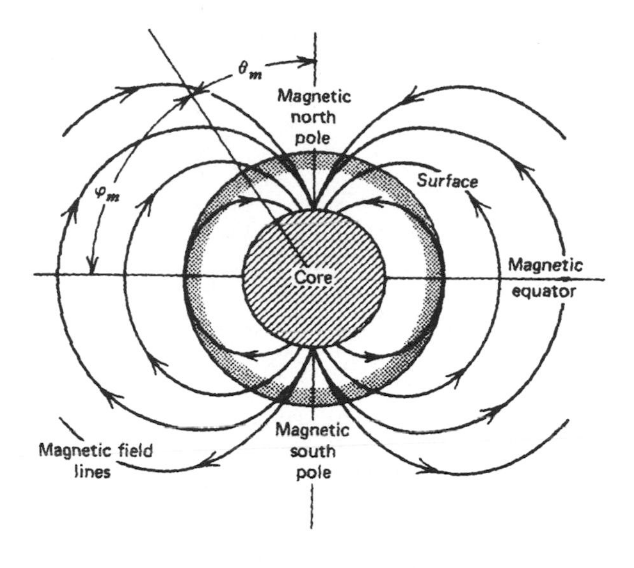

Fig. 7.7
Geomagnetic Lines of Force
(Geodynamics:1982-22)

representing the Fourier Analysis with respect to polar coordinates induced from a sphere. These force lines that depict the remanent magnetic field will vary in their influential magnitude through flux variations occurring from diverse orientations due to latitude and longitude, measured in micro-Teslas and graphically illustrated by plotted representations called isogonic lines (fig. 7.8). The numbers listed are in bearing compass numerics and will be identified in greater detail later (1991:152). The geofield has been shown to be in existence and of measurable quality. The next level of exposition is to determine if the field is an enclosure encompassing the planetary structure and, if so, the force lines qualify to act on our behalf and could be considered in any way acceptable in meeting the requirement of being identified as an atmospheric component.

The geofield structure illustrated by isogonic plot and compass bearing nomenclature used in surface navigation is sufficient in magnitude to influence a small metal object that has a magnetic dipole imparted to it. If this object is held in a certain horizontal position with the polarized metal free to move, the orientation of the poles imprinted in the metal piece will align

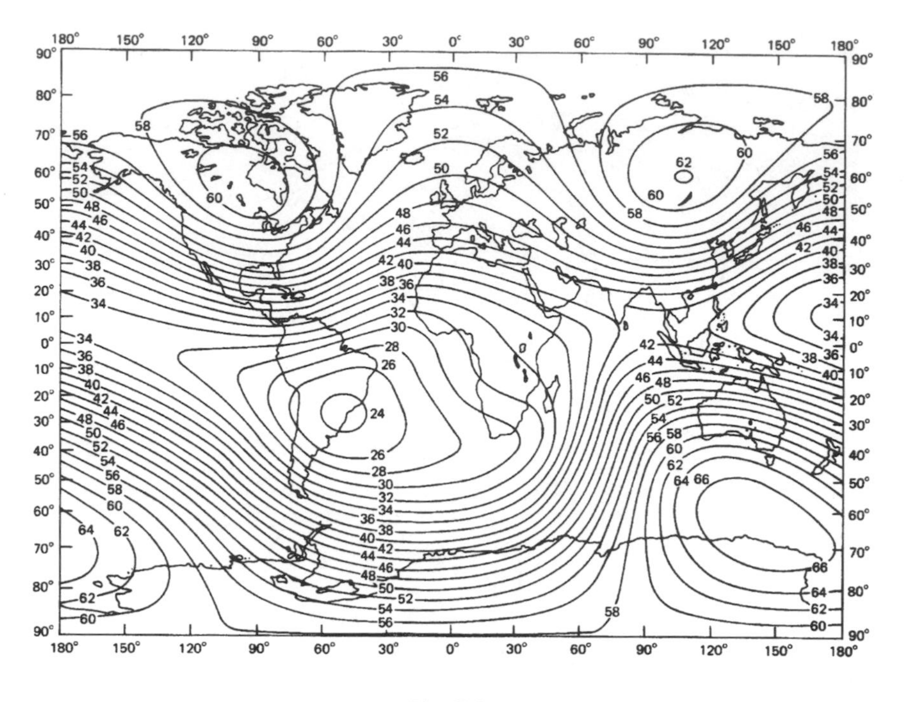

Fig. 7.8

Present Earth Geofield in Nanotesla's
(Geodynamics:1982-23)

along the lines of force as to be in parallel with and oriented with one end pointing in the direction of the Earth's north polar magnetic location. This has been demonstrated in past history in many aspects of our lives when traveling to a point that is distant enough as not to be visible terrestrially. Because Tesla measurements can be correlated from analyses, a representation of the field's strength can be plotted; however, the force lines themselves are invisible yet exist within the contained molecular atmosphere. Isogonic lines represent a charged particle influence on positive quanta, so they are therefore in motion from the magnetic south to the magnetic north pole. These nucleon particles surround us much like the molecular containments of oxygen, nitrogen, and water vapor. The charged particles, however, are flowing in a specific alignment and direction, whereas the molecules of the material atmosphere are scattered and omnidirectional.

On the surface the geofield is common place, yet it varies in magnitude. It is therefore most reasonable to determine if the density of magnitudes diminishes the farther from the surface the measurements are taken, rendering the field ineffectual. We seek to determine also that the more distant the force lines are from the terrestrial surface, they may at some point encompass the planet. The intent at this point is to determine whether or not the Earth's magnetic field is of an environment that may be applicable to a shield effect from higher-energy particles that produce severe and hazardous consequences and are in the spacial context of an encompassing sphere in addition to troposphere, stratosphere, and higher. This supplementary understanding will ascertain that high-energy particles impact our planet from the sun and determine whether or not the magnetic field is in some way a controlling influence over any detrimental inward-streaming solar plasma.

The issue of a shield must be considered within the context of its application. The geofield is a particle configuration, in the format of ionic motion from the southern magnetic pole to the northern magnetic pole. Quantum energies in this structure have unique properties beyond the allowance of ultraviolet wavelengths that penetrate and become absorbed by molecular ozone. Since ultraviolet is absorbed by molecular ozone, it must be seen that there is no shield effect to this particular wavelength from the force lines, thus allowing the radiation to penetrate into the troposphere for the purpose of creating monatomic and diatomic oxygen. An assumption therefore that the geofield must be, depending on the distance from the surface, coming into contact with other spectrum energies is our first step in shield explorations.

Plasma

Our planet is one among others that orbit a star. The star is our sun, and because it is a star, there are certain nucleon energies that are a result of internal reactions. The sun's mass contains a composition of three parts hydrogen to one part helium (1998:26), representing almost 99.9 percent of the total mass of our solar system. Therefore, because of this great mass, the pressure at the core is immense and effective in hydrogen conversion from the resultant incurred temperature. Temperatures of 27,000,000 degrees Farenheight cause the hydrogen nuclei in quantities of four to fuse

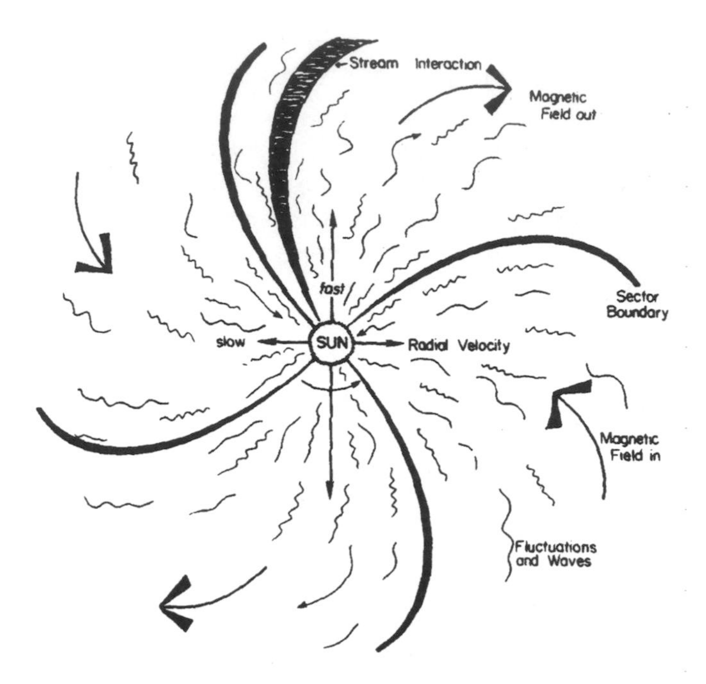

Fig. 7.9
Solar Wind (Equatorial Plane)
(Encyclopedia of Physics:1991-1141)

into one helium nucleus (Watters, 1995:39). At the photosphere of the sun, the generated energy from that conversion overcomes the gravitational field and proceeds outward in wavelengths containing visible and infrared radiation (.4 um–.7 um) with some shorter ultraviolet wavelengths. The imbalanced atoms of hydrogen and helium charges will propagate outward in a different manner from those of the radiation properties that are not fueled by the corona (Lerner and Trigg, 1991:1140). These atoms will be in a spiral format and at a lower velocity in the supersonic range, measurable at two or three solar radii (Merrill, McElhinny, and McFadden, 1998:383).

The corpuscular particles of free electron quanta in electromagnetic form, having been released through ionization by collision and high thermal temperatures within the sun's center, comprise the majority of the outward ionic radiation component. The two components (UV, visible

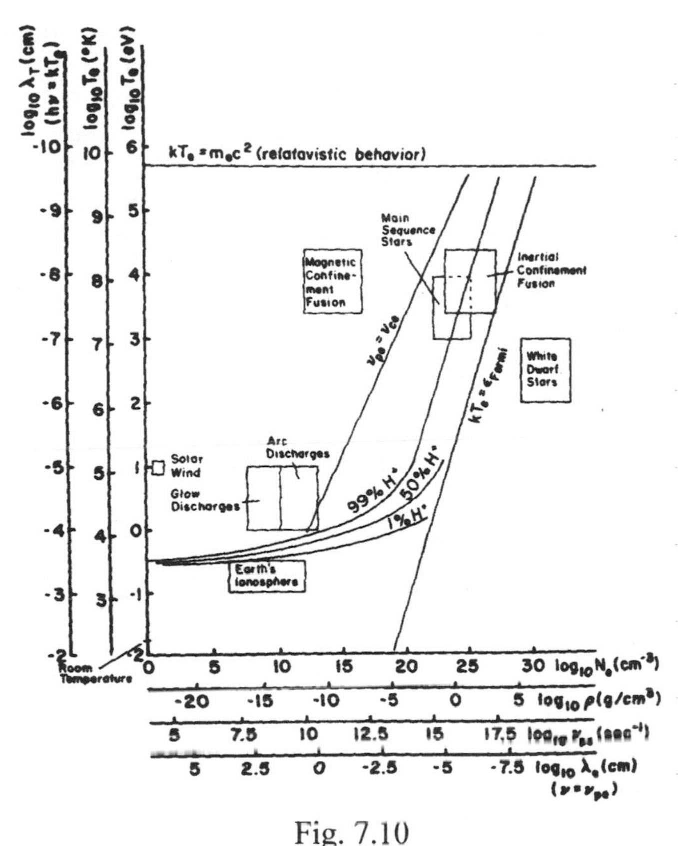

Fig. 7.10
Density (Ne, Free Electron) and Temperature (eV)
(Encyclopedia of Physics:1991-928)

infrared emissions and the slower medium of hydrogen and helium ions) can be seen in figure 7.9, representing this propagating structure.

The definition of this dispersion requires a majority of free electrons for electrical conductivity and the kinetic energy of the free electron must be greater than the specified value for its density (value will increase with increasing density). When these criteria are met, the term *plasma* may be applied (1991:928) (fig. 7.10). This plasma and the associated sun's magnetic field that is outwardly scattered is identified as the solar wind and becomes our focus in an analysis through which the Earth's geofield comes into contact with it.

Observations strongly indicate that at the sun's surface a curved falling effect can be seen in visible light. This presupposes that it had to ascend from the surface and that not all the visible matter represents the totality of outpouring energies. Using this as a foundation, an assumption can be made that there is a portion of the surface outgassing that is invisible and the invisible energy must be in association with the hydrogen and helium charges that can be measured from vast distances beyond the sun itself. If these cases are indeed true, then the Earth itself would be in the trajectory's path (Lang and Gingrich, 1979:128) and that an interaction between the geofield and solar wind would exist.

Magnetic Shield

Earlier in this chapter, figure 7.1 illustrated that irradiance can be measured at the top of our atmosphere (the molecular atmosphere) and that the molecules and atomic structures comprising that atmosphere will react to specific wavelengths of charged particle energies. With this in mind and the fact that a geofield of electromagnetic properties is involved, further exploration of that electromagnetic interchange can be undertaken.

A shield protects or defends from something. Our search is to determine if *protection* is an accurate description of the interactivity between free electrons in motion as a constituent of the solar wind and the ion motion that is represented by the Earth's geofield. Essentially, when free electrons are in motion in any capacity, that is generally accepted as an electrical current. Here the motion of free electrons within the plasma eventually impinges upon other particles that are field-aligned and are also in motion. Since geofield field lines are aligned with electron flow (current) in the direction from south (negative) to north (positive), that charge alignment will have an influence upon the ion particles arriving from the sun.

Solar ionic plasma contacts the Earth's geofield at certain angular vectors because of the Archimedes Spiral induced by the sun's rotation. (Figure 7.9 identifies this spiral nicely.) Because of this fact, the ion bombardment initiates a particular effect when proximity to the field boundary wall occurs. Because electron particles radiating outward within the plasma field are at some velocity and have spin, they will become magnetically induced when under an influence of a field line consistent with that

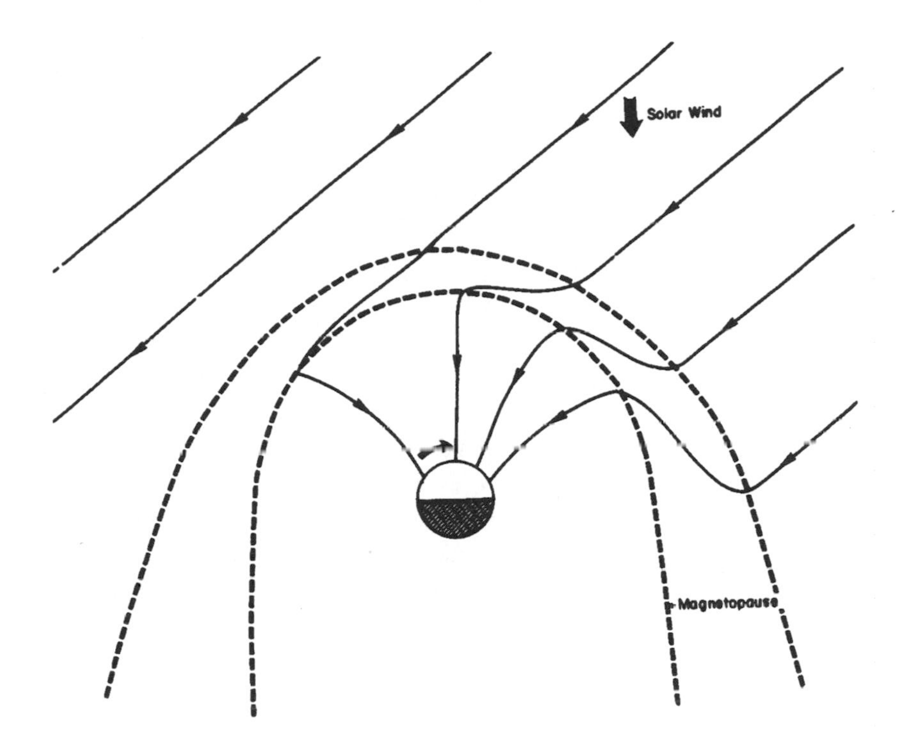

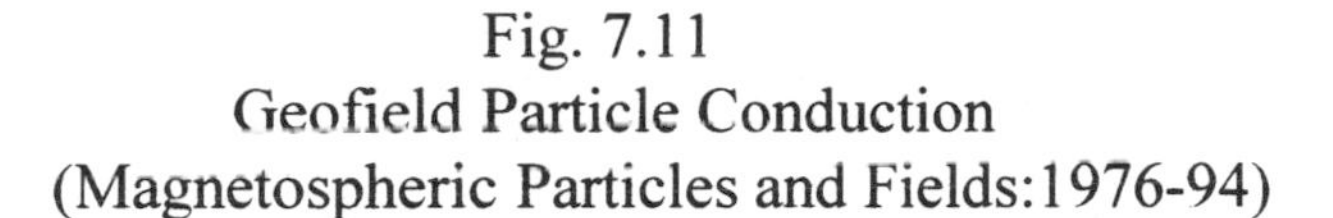
Fig. 7.11
Geofield Particle Conduction
(Magnetospheric Particles and Fields:1976-94)

particle's motion. Depending upon the impingement vector, some of the solar plasma will couple with the geofield lines precisely because of their spin and velocity. However, not all of the solar plasma will couple with the geofield lines, and depending upon the trajectory vector, some plasma will be directed around the Earth and some reflected back to space (fig. 7.11). In view of the focus to determine geofield properties that are influential in the detour of harmful plasma characteristics, a closer geofield analysis on boundary impingement vectors identifies properties very important in solar particle displacement.

The geofield itself is an obstacle to the solar wind, generating several unique reactions that define its presence as a bowshock to the plasma constituents of protons, electrons, and other solar particles. There is a certain

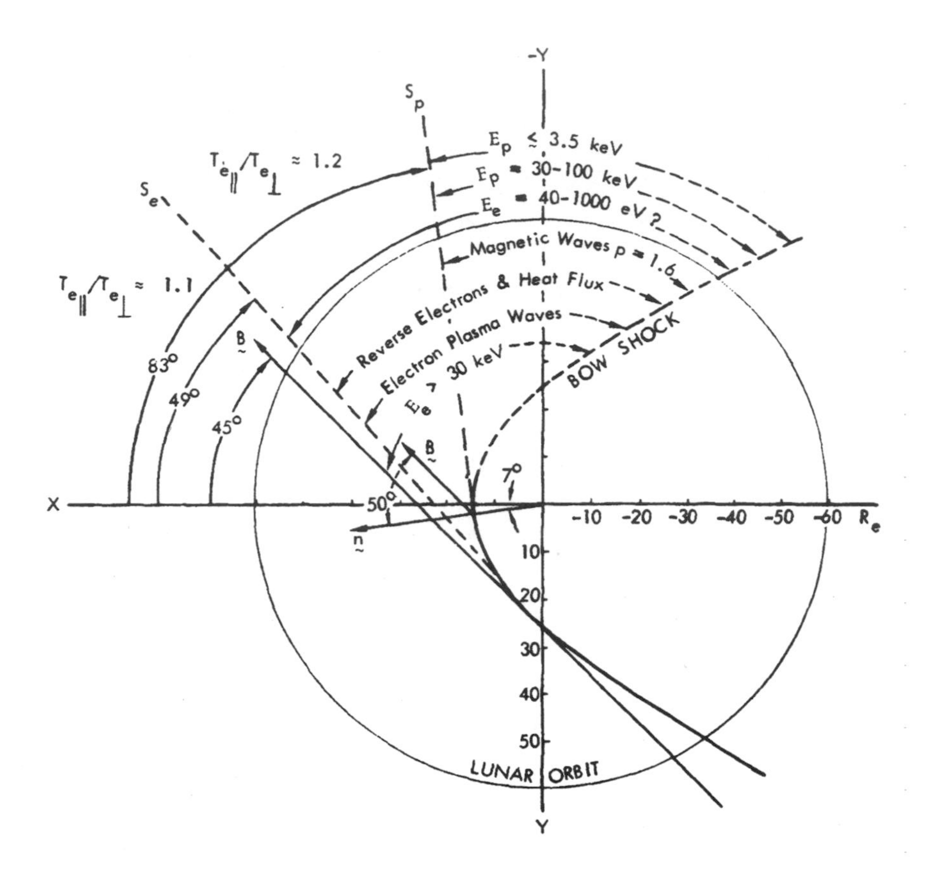

Fig. 7.12

Geofield Shock Boundary and Phenomenology

(Magnetospheric Particles and Fields:1976-14)

vector relationship that provides for these particle in their escapement from field line coupling and can be illustrated in figure 7.12. With inspection to this pictorial, there are a series of dashed lines labeled *Se* and *Sp,* respectively, and these define the angular umbrella between forty-nine degrees and eighty-three degrees measured from axis X; it is this sector that will be discussed.

Prior to reaching the Earth's geofield, solar particles within the wind

field (plasma) will invariably collide with each other. Howeve[illegible]
uation the collisions require encounters and because of the great [illegible]
between the particles, a collisionless plasma definition of the solar w[illegible]
in effect (1976:121). It is this collisionless factor of the wind that will [illegible]
low its entrapment and eventual containment within the geofield lines of
force. This will be explored in greater detail later.

Solar particles have velocity and represent a charge (quanta); they will emit energy in the form of an influence on other energies around them; this energy about the particle under velocity is an electromagnetic field (1990:111). Taken further, a charged particle with velocity in a magnetic field will then be subject to outside influences upon that field in such a manner that the influence being imparted to that particle through its magnetic field will be in its eventual direction and is a function of whether the direction is parallel or perpendicular to the Earth's geofield lines of force. The original direction and velocity is a result of solar ejections, however; some emissions are faster than others, and as a result of this overtaking plasma the outward spiral will tighten increasing particle velocities (1997:124). Of course the more parallel the direction to the geofield's lines of force the particles are, the less the chance of particle displacement (1997:123). Although the solar wind contours will change because of the differences in velocities from solar ejections, flares, et cetera, the magnetic fields generated by individual particle spin and velocities will not be altered. This particle magnetic stability is called a frozen-in field and since plasma (particles) are a conductor, they will generate an electrical current because of that field stability (1997:121).

Upon contact with the Earth's geofield, the preceding conditions react to the extent that the particles, their velocity, and angular relationships as well as their own frozen-in magnetic field all play an important role in that encounter. When contact is made, the geofield will deform (fig. 7.11) due to magnetic pressure and, because of the interrelationship between fields, the plasma will exhibit an increase in energies termed *heating*. It is this heating that determines particle coupling to the geofield line and also their potential confinement. In addition, the particle's electric field will cause velocity accelerations if the geofield lines and particle electric field are parallel. It is this situation that causes a de-coupling effect that eventually removes the trapped particle and allows it to continue away (1976:126). Also in this discussion are that particle energies are also coincident with periods of solar wind velocities and that part of their energy increase can be attributed to that fact that the closer they get to the Earth, the

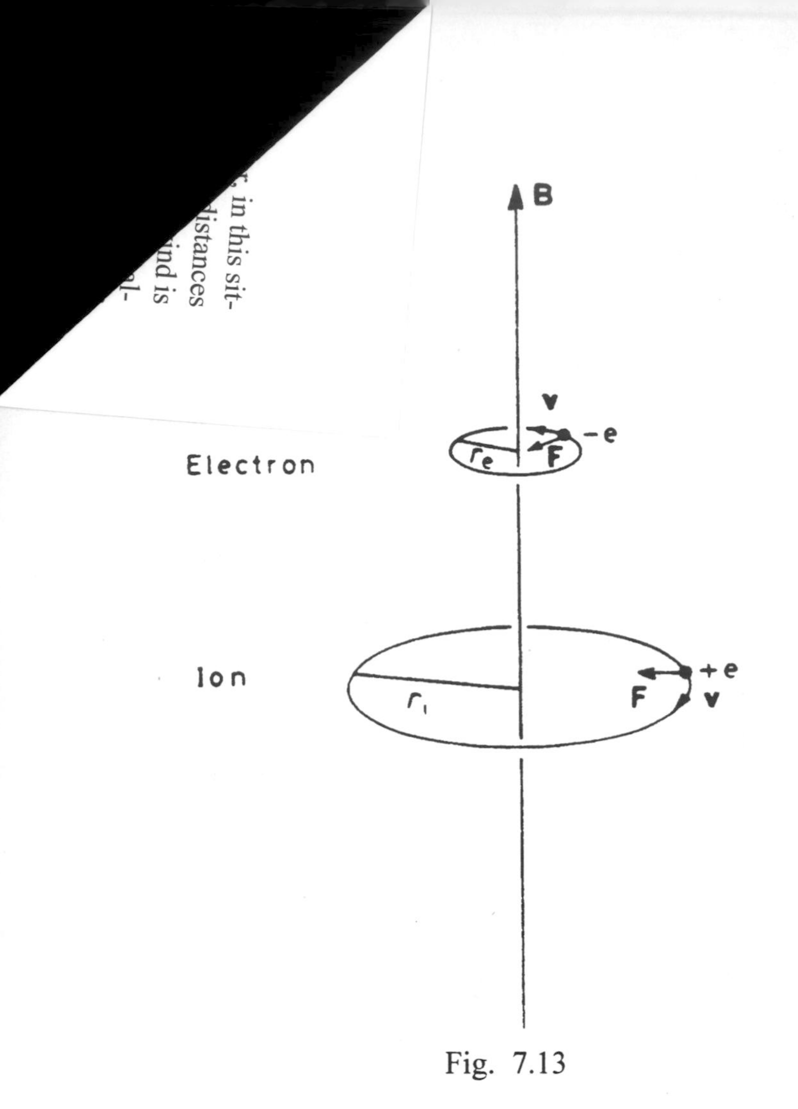

Fig. 7.13

Betatronic Acceleration (gyrofrequency) about Geofield Lines
(Solar-Terrestrial Environment:1992-15)

shorter the field lines become and, therefore, the higher the field intensity will be (Hargreaves, 1992:187).

The physics behind such increases involve the rotation of the particle about the field line (gyrofrequency) to the extent that as the flux density of the field increases, so will the rotation frequency in like proportion (betatron acceleration). Since the angular momentum is conserved in this process, the kinetic energy of the particle will increase (Hargreaves, 1992:15) (fig. 7.13).

The particles that travel along the closed field lines of the Earth will extend their motion along those field lines several earth radii into space. These are considered to be trapped particles and consist of protons and electrons that are referred to being within the Van Allen Radiation Belt encompassing geofield lines. These particles are considered to be the most energetic and in some apportion comprise the ionization factor of our upper atmosphere when they leave the trapped lines of that region (Hargreaves, 1992:164) (fig. 7.14).

Particles that do not couple to the geofield lines upon first contact with the magnetosphere are diffused back upstream mainly because of their impingement (impact pitch) angle, which in figure 7.12 is located between the dashed lines of Se and Sp, or roughly between forty-nine and eighty-three degrees. The illustration also shows that the energies of plasma electrons and protons increase at the earliest encounter and elevate the mean energies between thirteen and fifty electron volts (1976:14).

Particles from the solar wind plasma will in their coupling to a geofield line flow several ways. They typically travel around the Earth, and where the geofield is somewhat open particle entry is possible. This is especially evident at the north polar region, where particle entry can be seen visually as the aurora. Interestingly enough, the geofield does provide for plasma protection due to the magnetic relationship between particle fields and the magnetosphere coupling particles to the polar region. In addition, depending upon the contact angle with the plasma and lines of force, this will deflect some particles back to space. The invisible part of the geofield establishes a form of protective barrier from the high-energy solar wind; even so, as some of the solar wind is deflected around the outer geofield, those particles along the closed lines will eventually slow their velocities and come into contact with the lower atmospheric structures. Because of this contact, we experience this as a form of electron storm, which in some cases will affect our weather. From spacecraft investigations it does appear that the sun's plasma discharges do affect us because their intensity, properties, and pitch angle are disruptive, especially in radio communications and electrical phenomena. But, it is precisely because of the interaction between the magnetic properties and the geofield and solar plasma that the intensities from such energy bombardments are mitigated.

At this stage of the expositional investigation into the Earth's atmo- and magnetospheres, there has yet not been any work done to learn how to provide this information on a daily basis. We can experience rain, snow,

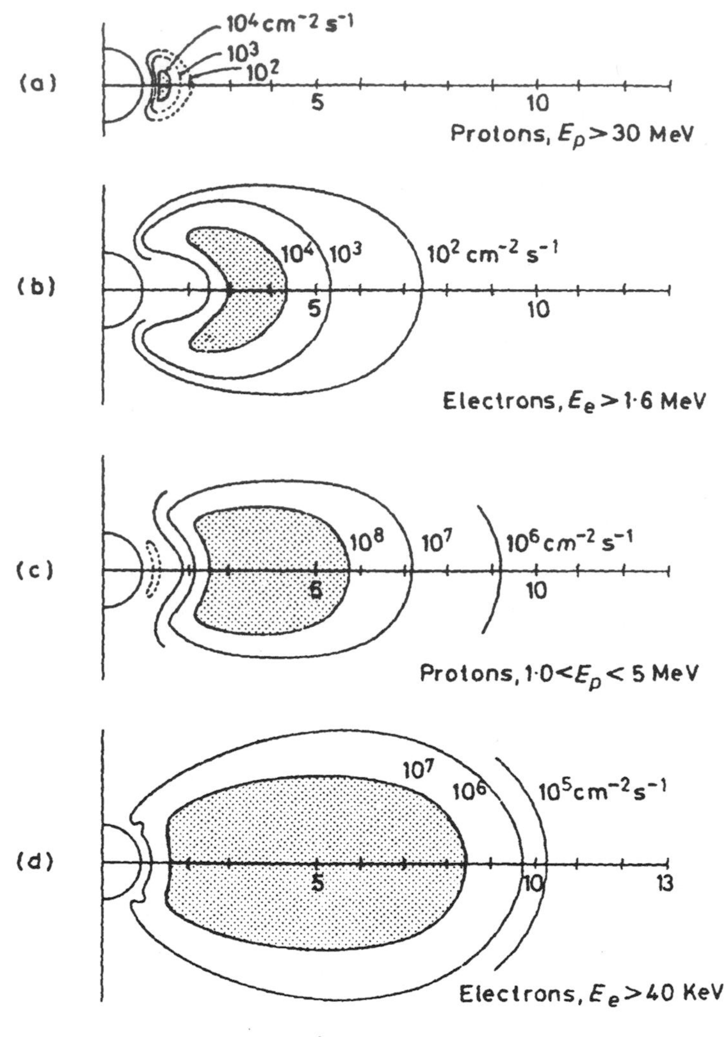

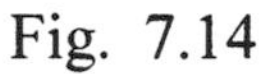

Fig. 7.14

Van Allen Radiation Belt vs Particle Energies
(Solar-Terrestrial Environment:1992-1·88)
Earth Radii @ a,b,c,d

and fog. We can see the clouds and because of them know that a particular frontal system is moving in our direction. However, the presentation of these properties has not been explored. The following chapters will now concentrate on that presentation, using previous knowledge learned and its application to atmospheric charting.

8

Creative Transformation of Thought

In the Beginning

At one time we could only look up and guess what was going to happen, and that situation was not long ago. Believe it or not, there was a belief that the weather and climate in general were controlled by gods. Church bells would be rung during periods of inclement weather to ward off what was thought to be the cause, as it pertained to theological evil spirits. Weather situations were for the most part relegated to local areas, understandably, until Paramenides and Aristotle in 500 B.C. began to describe the fundamental differences between climatic patterns and associate behaviors reported from various parts of the known world (Fleming, 1990:2). In the explanations given, the height of the sun over the horizon was considered the primary cause for latitudinal variations of cold and warm, wet and dry seasons, which they termed *climate.* This was a term that meant "inclination," which suited their reasoning nicely.

This explanation was not the most accurate, because in the winter of 1607–08 the weather was extremely colder than usual in the European regions from which they originated. Weather speculation as to its creation and origins was not confined to the agricultural communities. In fact, the medical representation of Salisbury, Massachusetts, 1822, in particular, theorized that the weather was linked to the health of patients (Fleming, 1990:5). Many other speculative theories abounded prior to the acquisition of a systematic methodology of collecting data.

In 1591, the snowflake was speculated as having either six or eight sides; this was presented by Thomas Harriot, who was the scientific adviser to the Roanoke settlement in Virginia in 1585–86 (Allaby, 1998:72). Atmospheric convection in the rising and falling at the Equator and polar regions, respectively, was introduced as the fundamental reason behind wind currents. Not only was George Hadley (1685–1786) instrumental in the formation of convection cell theory; he was also most intuitive in the reasoning that the convective nature in both north and southern latitudes was a byproduct that stemmed from the Earth's rotation.

This was confirmed in 1835 by Gaspard de Coriolis (1792–1843) and later refined to submit the rationale that atmospheric flow will be clockwise about the center of a high-pressure area and counterclockwise about the center of a low-pressure area. We now recognize this as the Coriolis Force. Later, in 1857, this phenomenon was clarified further by Buys Ballot (1817–90) through observations regarding this fact, although the theory behind this effect had been presented by William Ferrel (1817–91) in 1857 (Allaby, 1998:91).

All these men were relegated to theoretical postulations that were essentially based upon their observation at the surface level and were not primarily derived through the systematic collection of multiple data locations either regionally, terrestrially, or by vertical atmospheric sampling. Eventually, the recording of data that was to be the cornerstone of multilocal sampling would come from the migration and settlements that formed the expansion of our nation.

Obviously, one cannot expand into territory that was not part of this nation without some form of ownership rights, either by defense and eventual military conquering or through an acquisition of some sort. The West lay open; however, to "get west" required travel through the areas that lay east of it. As it turns out, Napoléon was the key. The current states of Louisiana, Arkansas, Missouri, Iowa, North and South Dakota, Nebraska, Oklahoma, sections of Montana, Wyoming, and Colorado, and some of Kansas were sold for the sum of $15 million to the United States under Pres. Thomas Jefferson. This part of our United States was known as the Louisiana Purchase (Laskins, 1996:95). It was with this purchase in 1803 (1982:223) that Jefferson effectively doubled the nation's size and thereby created the necessity that some form of exploration into this newly acquired territory must take place, for President Jefferson in actuality had no idea what he had purchased any more than Napoléon had of what he had just sold. It was at this point, on May 14, 1804, that a captain in the army by the name of Meriwether Lewis and a lieutenant, William Clark, also in the army, began the undertaking of an exploration of this vast new land. Because a diary of events and "conditions" was vital to provide the president with exploratory reports, the chronicling of the events that were to shape our meteorological future were part of this undertaking.

In the two and one-half years of the Lewis and Clark exploration, some observations were quite interesting. First, lightning was observed to be more frequent in these territories than had previously been observed in the eastern Atlantic region. Winters were tabulated by their severity, recording below zero temperatures in the Dakotas, where the team spent the 1805 winter. Observations were made regarding the gradient force winds being of greater violence than in the coastal Atlantic regions. A reason for this was given as the absence of trees. Additional observations on the rapidity of thunderstorms not only in their formation but also in their water content were chronicled when the team left the plains and started their ascent into the Rockies. All the varieties of climate were dutifully recorded without personal comment; only the observations at their locale were written.

Before the Lewis and Clark party returned from their exploration, another team of twenty-three Caucasian men and fifty-one Indians (Laskin, 1996:99) set forth to catalog the region from the Missouri River to the Arkansas River and follow that river to its source. Since their method of writing was not in a structured form of military "just the facts" style, he added more information that apparently offered insight into causation of topography. He used words to describe topography such as *arid, hot, parched,* and of course *desert.* This had a special meaning to an agrarian society that was not aware of the middle state climate, and they (East Coast Atlantic) could only envision that the middle part of this great purchase was a vast Sahara and of course of no value.

Although some accounts dissuaded many from the arduous travel "out west," some thirty thousand who did make the journey in 1849 searching for the Californian gold did chronicle their daily travels and included the climate that was encountered along the way. This was an important step in predictabilities that may be presented to individuals planning for the eventual climate, which may be unsuitable along the way, depending upon the season and locale one must traverse. In this manner, descriptions were employed for the first time to describe the features in words such as *blizzard,* which was used for the first time in an Iowa newspaper to describe the intensity of the snowstorm that befell that populace on March 14, 1870. It was here that a descriptive meaning of temperatures that held steady below twenty degrees Fahrenheit with a wind constant at thirty-five miles per hour was established (Laskin, 1996:110).

It was here, at this junction in our nation's historical cxpansion and economic development, that the writings of the new territory's climate needed to have some form of "real-time" representation concerning meteorological events. Correspondence from many miles away takes weeks to reach East Coast papers, and then several more days were needed to have the information printed for readership. This time of the essence, so to speak, was a crucial point in the establishment of any meaningful presentation regarding the events out west and the possibility of those events coming east. The only appreciable source of expected climate conditions that could be of any use was *Poor Richard's Almanac,* which was first published in 1732. A newer and more advanced method of "this instant" type weather forecasting was needed, and with that need would come the technology to measure and record the atmospheric behaviors that seemed, at least, to originate out west.

Instrumentation

Observations are required of any data correlation and record. However in view of this, observations using just the eye and touch (wet, dry, hot, et cetera), utilize our natural sensitivity as the origin of a standard to enhance that perspective, and something else must be needed. That need will take the place of some form of measurement, which suggests a numerical scale (1966:8). By that statement, I mean that the events being observed must conform to a prior set of perceived values and that a comparison of accepted values weighed beside the newly obtained observation causes the rendering of a conclusion that will be founded upon differences or equalities between them. As an example of this, we can say that the temperature outside is hot. First, we need a definition of the situation regarding the word temperature to render a conclusion that is suitable to utilize the word hot. Second, circumstances regarding the area of thermal properties must fit the standard to which the definition of *hot* implies, and third, a set of values that identify a criterion from which the word *hot* is brought about, as the form of a conclusion from those comparisons between a thermal definition and an appropriate observation or collection of applicable parameters.

Nature itself produces a variety of thermometers. Two are an ant's speed as it moves and the chirps of a cricket. As the cricket chirps, add the numeral 40 to the count after fourteen seconds. Know, therefore, there is always a variable due to shaded areas in which the cricket may be in, so . . .

the number that represents the temperature may be slightly different from where you may be in reference to the cricket's locale. In the varieties of flowers, a rhododendron's leaf will drop as the temperature falls and, the *Eponymous radicans*'s color will vary in the shade of red progressing to deeper reddish colors, eventually to the color of black (1988:25).

In 1593, an invention termed *thermoscope* was created by Galileo (1564–1642); however, because this invention operated on the fact that air expands, the representation it presented was inaccurate, due to the physical fact that any volume of air will vary due to atmospheric pressure around it. This, however, was still of great significance, because it was fundamental to the premise of molecular expansion (increased activity) insofar as the higher the surrounding temperature became, the more the molecules translated that energy into a higher state of motion, thereupon causing the warmed liquid to rise, or fall if cooled (contract), within confines of a tube. The reality that optically observed physical variations and subsequent quantification produced by a physical mechanism in response to the presence of an invisible force demonstrated conclusively that there was an association between meteorological events and devices that would react to them. Obviously, this invention was followed with interest.

Not long after Galileo's death in 1642, the liquid was altered from water to wine and given its observation color through the use of dried kermes. The top of Galileo's thermometer was designed to be open and now was sealed. Later, another change in the liquid constituent was made from water to mercury by Athanasius Kircher, in 1620. At least here, the invention of a device that presented changes to what we had perceived as differences in heat and cold, was the first step in the formation of solid references to quantify these changes and to provide data that could be used in identifying those variabilities against a terrestrial scale. Later, that historical database could be used to identify regional expectations over time and would be useful in the assistance of rudimentary predictabilities.

In 1714, Gabriel Daniel Fahrenheit (1686–1736) improved the existing thermometer to have a greater accuracy over a variety of thermal extremes. One of the previous concerns with alcohol-filled tubes was the low boiling point that these types of measurement devices had, and as a consequence of this fact, they were very inoperable at high temperatures. Instead, he used mercury for the liquid matter and, in concert with that new substance, created a scale that would encompass the high temperature of boiling (212 degrees Fahrenheit) and the freezing point of pure water (32 degrees Fahrenheit). Because of this advancement, both the mercury thermometer and the Fahrenheit scale are still in use today (1998:79).

An additional numerical scale was also introduced to provide a different scaling factor. It was in 1742 that this alternative to the Fahrenheit nomenclature was introduced by the director of Sweden's first observatory, Anders Celsius (1701–44). The Celsius scale prior to 1948 was recognized as centigrade; however, in 1948 centigrade was abandoned in favor of the inventor's name. Through the work of Celsius, the freezing point of water was identified by the decimal numeric of 0 degrees (Celsius) and, conversely, the boiling point at 100 degrees (Celsius).

Although the thermal properties of our molecular atmosphere were of interest and were suitably covered by the thermometer's creation, the fact that molecules under compression will produce heat is also of vital concern. The question then is asked, "How do we know that the molecules within our atmosphere are being compressed, or allowed to expand?" Does the temperature itself provide that information? Is there some way that temperature and atmospheric densities are related? Does one cause the other?

Since the Atomists had defined our atmosphere as a composition of indivisible atomic structures, then perhaps there was a way to measure the sum of these structures within a certain locale and, in addition, whether the amount or density of structures varied accordingly with geographical changes.

It would appear that Galileo's thinking and thermoscope demonstration was not lost on his assistant. Shortly after Galileo's death in 1642, Evangelista Torricelli modified his mentor's thinking by advancing the theory of liquid displacement. Just as the mercury within a confined tube will expand when heated and contract when cooled, i.e., rise and fall, by filling a tube with mercury, enclosing one end, and placing the other open end in a volume of mercury also, as atmospheric density presses upon or retracts from the liquid surface area, the molecules will respond by increased collisions against the dishes confines. This activity then causes a liquid response by its progression of either up or down within the tube until the tube's liquid column is equal to the atmospheric weighting outside, represented upon the dish's liquid surface (Lutgens and Tarbuck, 2001:158). Although the rising and falling of mercury levels with respect to the thermometer and barometer are indeed similar to each other, in the case of the thermometer the rise in level is through thermal molecular activity initiated by the direct result of temperature involvement. Whereas the mercury tube has the open end submerged within a dish of like substance and the other tube end closed and containing a vacuum, the rising and/or falling of the tube liquid is observed through the effect of collision displacement, previously stated as atmospheric influences acting upon the dish liquid's broader surface (fig. 5.9). The significance of this theory and its application within the discipline of molecular dynamics are illustrated in Chapter 5 by this physical law.

An additional enhancement to the barometer was implemented by replacing the liquid. This change defined an aneroid barometer (Lutgens and Tarbuck, 2001:159). *Aneroid* means "without liquid." Replacing the liquid was a partially evacuated metal chamber that responded to atmospheric weighting through a change in its dimensions (fig. 8.1). The movement of chamber dimensions is then attached to an indicator imposed upon a numeric scale, which represents those fluctuations of atmospheric pressure as defined more accurately in Chapter 6.

Data Communication

With the two forms of devices previously defined, numeric observations could now be recorded in a manner that would support the foundation of a rudimentary database. What was lacking, however, was the ability to transmit this data to a distant location that provided assessments in real time. For example: Previous measurements from the instruments mentioned provided records at that location only, and any transmission of that data would be in some form of paper communication to another destination requiring a passage of time for delivery. Any place other than the actual location of the device was therefore receiving historical data as to what had previously occurred and not what was happening. As a consequence of this, since weather conditions predominately move from either the westerly regions to the east or from the south to the northeast or north to southeast, any analysis of patterns that could in some way provide a meaningful forecast of eventual East Coast climate was ineffectual, much less from any point of western origin to any point east of that affected region.

The principles of electromagnetism and electricity changed this situation through the observations of Gilbert, who in the sixteenth century discovered that certain substances contain properties that possess an attractive force that he called electrics. Later, in the early 1800s, Hans

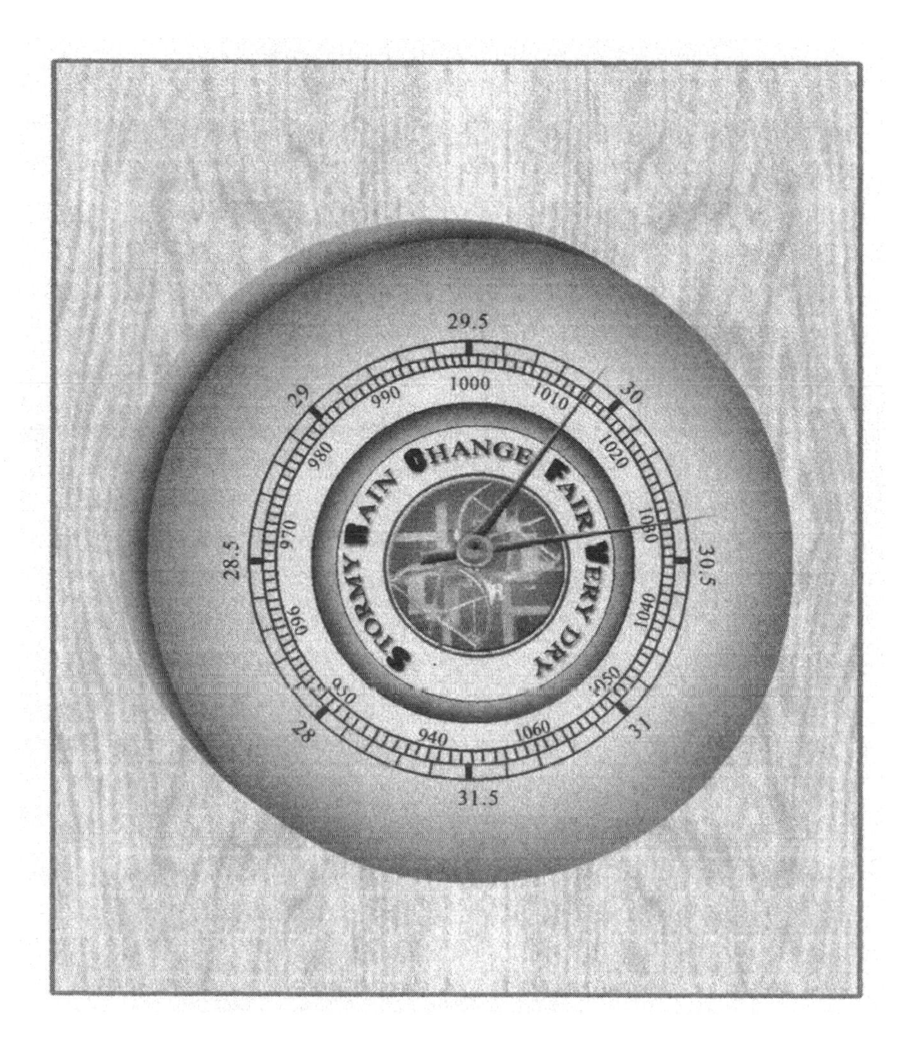

Fig. 8.1
Aneroid Barometer
(Mendes-Hussey Graphics)

Christian Oersted (1777–1851) expanded that thinking when he developed the theory showing a relationship (in approximately 1820) between electrons flowing (current) in a copper wire and an invisible force that was generated and observed by the deflection of a metallic needle (magnetism) when the current and wire were placed in proximity to each other (Williams, 1999:55). Because of this discovery and subsequent work in that field, the quantity of magnetism called oersted is in his honor.

The principles of an electric current flowing within the confines of a copper wire were modified by Sturgeon when he made straight copper wire into a coil and reapplied electrical current, thus creating the first electromagnet. Michael Faraday (1791–1867) improved this theory when he discovered that an electrical current could be achieved by the process of moving a magnet through a wire coil. In this way, he demonstrated that fluctuations of a magnetic field will produce electric currents and voltages (Branson, 1967:3). Through this discovery eventually came the electric generator.

Joseph Henry (1807–78) (fig. 8.2) used the electromagnet in a very unique way. He was one of the most important scientists in that day, who actually created the fundamentals of electromagnetic construction. This was of course most useful in transmitting electrical energy over a conductance that converted that energy into a mechanical function. Just prior to Mr. Henry's work was the discovery by a Danish physicist named Hans Christian Oersted in 1820, which proved the existence of an invisible force at work when the deflection of a magnetized needle was observed as a result of passing an electrical current through a wire. Although this was of course important in some form of remote generating influence, an additional step was required, and that came from William Sturgeon in England. Mr. Sturgeon took an iron piece that was shaped in a horseshoe and wrapped wire in several loose turns about it. He found that the iron became magnetized when an electrical current was passed through the wire winding and became nonmagnetized when the electrical current was turned off. This was the electromagnet that Mr. Henry would eventually improve upon.

In 1928 Mr. Henry demonstrated that by additional windings of some four hundred feet of insulated wire, wrapped in a tighter configuration about the horseshoe-shaped iron, he could create a most powerful magnetic influence. This was a great breakthrough in the understanding of magnetic induction, and thus the unit of that force that now bears the name Henry. His findings were reported in Benjamin Sillman's *American Journal of Science* in January of 1831. This is not to say that Mr. Henry was the

Fig. 8.2
Joseph Henry, 1797–1878
(National Portrait Gallery, Smithsonian Institution)

first to invent the electrical apparatus that transmitted some form of communication in a code that would be deciphered at the other end. That invention had been in existence almost 100 years. Henry did, however, effectively create a simple mechanism whereby a sound would be produced through his electromagnet applications. In 1831 he demonstrated to his class at Princeton University this effect. He placed a permanent magnet between the poles of his horseshoe magnet and by reversing the electrical current through the horseshoe magnet caused the permanent magnet to be either repelled or attracted to either pole end. When the permanent magnet was influenced to move, it would tap a bell. This particular experiment was carried out with over one mile of copper wire strung between campus buildings. In addition, Mr. Henry was the first to demonstrate at Princeton that by applying electrical energy to greater winding configurations metal loads could be lifted. When the electrical energy was turned off, the metal loads would drop to the floor. This was the first real demonstration of a relay, although that would be refined at a later time.

The actual transmission of code was accomplished in 1832 by Shilling when he used six wires to send an electrical current through which seven magnetized needles were controlled by a rudimentary keyboard of colored keys (fig. 8.3). Depending upon the key configuration, either polarity could be sent down the wires, affecting the magnetized needles by causing them to reorient themselves with relation to the current flow within the wire. The orientation of the needles was then a representation of the transmission word. This was in essence the first control of some form of distant communication. Although this was in use in Europe in 1832, the military or commercial usage was not pursued with great interest.

Magnetized needles were used for many years without much variation to their application until the generation of a single wire configuration that would in some way support bidirectional transmissions. Although needle response was effective, the invention of the electromagnet provided the last piece of eventual mechanical solutions through remote means, and it was this invention of inductive force influences by way of propagative electron current that Mr. Henry should have been given credit for and was not. Later publications were to make only a passing reference to Mr. Henry and not offer the actual history of his work, which would have clearly influenced any patent outcomes for another individual granted in 1846.

The advancement of the single-wire application was enhanced by Prof. Samuel Finley Breese Morse and a Prof. Leonard Gale, who was a chemist and colleague, in 1832. Morse's signals were so weak that they were not appreciably reliable over twelve meters of wire in any efforts to elicit a remote response from the transmission of electrical current in 1836 (Coe, 1993:29). Professor Gale was, however, familiar with Mr. Henry's electromagnetic induction theories that were already published in the 1831 paper and adapted that thinking to Mr. Morse's application. The inductive coil turns of fine copper wire to amplify the field necessary to provide an amplified force through the electromagnet at the receiving end was done by Mr. Morse, based upon Professor Gale's recommendation through Mr. Henry's work. This amplification of induction through the increased number of coil turns would then be able to provide the necessary influence to obtain a detectable mechanical response (2001:52). One piece was missing, that being a source of suitable electrical current necessary to generate the amplified magnetic field at the improved receiving coil. With Professor Gale's knowledge of chemistry, the battery cell that was used as a current source was also increased from Mr. Morse's one battery cell to twenty cells, or thirty-nine volts. This combination, from an increased inductive capacity through the additional 100 turns of copper wire at the receiver's coil and additional battery cells to boost the voltage and current, respectively, was the impetus that initiated the communication revolution.

With additional refinements, the signal capacity for sufficient inductive observations by a distant electromagnet was increased to ten miles, and the patent was applied for on October 3, 1837. On May 14, 1844, the first words sent over the single wire, "What hath God Wrought," were transmitted between Washington and Baltimore by approval of the U.S. Congress with funding of $30,000, all contained within a bill dated March 3, 1843 (fig. 8.4) (Coe, 1993:32).

Here at last was an apparatus that would transmit through the presence of an electrical current carried by the conductance of copper wire, a form of translatable code that was derived from the duration that the electrical current was present. This form of language was identified as American Morse Code, so termed after the creator of it (fig. 8.5). Although the telegraph was widely accepted for many reasons, essentially political, for whatever reason, it was established that a descriptive language could be transmitted by the generation of electrical durations called pulses and, most important, that this language could carry meaning was essential in prompting considerations for enhancement of distance capabilities.

Any consideration for an appreciable distance had to overcome the physical fact that the copper wire presented a resistance to the electrical current over distance. This resistance within the wire would cause a significant less-

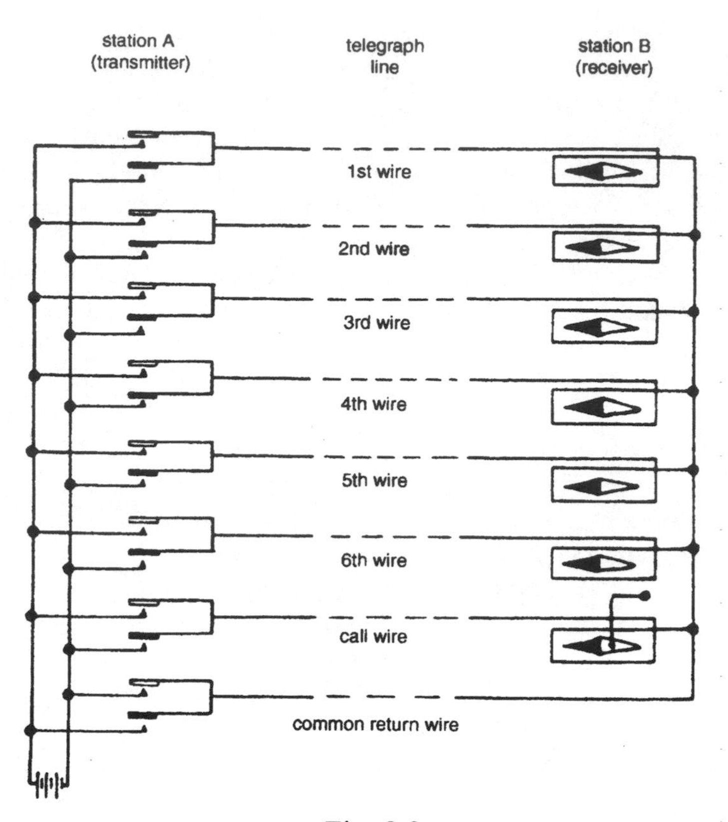

Fig. 8.3

Schilling's Six Needle Circuitry
(History of Telegraph:2001-29)

ening of the potential electrical current received at the other end, and this degradation in value was sufficient to be less than that required to create an inductive field necessary to activate the receiving electromagnet into a response. This distance before some form of current amplification was required was about thirty-two kilometers. Obviously, something was needed so that information could be sent over greater distances if the West Coast, middle states, or East Coast could transmit real-time observations or reporting.

This problem was addressed by an English surgeon in an attempt to surmount this barrier. Previously, considerations using greater and greater

Bill Passed by Congress Enabling Morse to Construct the First Telegraph Line, Between Washington and Baltimore

A Bill to Test the Practicality of Establishing a System of Electro-Magnetic Telegraphs by the United States. **March 3, 1843**

Be it enacted by the Senate and House of Representatives of the United States in Congress assembled, **That the sum of thirty thousand dollars be, and is hereby appropriated, out of any moneys in the treasury not otherwise appropriated, for testing the capacity and usefulness of the system of electro magnetic telegraphs invented by Samuel F. B. Morse, of New York, for the use of the Government of the United States, by constructing a line of said electro magnetic telegraphs, under the superintendence of Professor Samuel F. B. Morse, of such length and between such points as shall fully test its practicability and utility; and that the same shall be expended under the direction of the Postmaster General, upon the application of said Morse.**

SEC. 2 ***And be it further enacted,*** **That the Postmaster General be, and he is hereby, authorized to pay, out of the aforesaid thirty thousand dollars, to the said Samuel F. B. Morse, and the persons employed under him, such sums of money as he may deem to be a fair compensation for the services of the said Samuel F. B. Morse and the persons employed under him, in constructing and in superintending the construction of the said line of telegraphs authorized by this bill.**

Fig. 8.4

US Congress: Electromagnetic Telegraph Practicality Bill
The Telegraph: A History of Morse,s Invention
and its Predecessors in the United States copyright 1993
Lewis Coe by permission of McFarland & Company, Inc.
Box 611, Jefferson NC 28640

(a) AMERICAN MORSE CODE

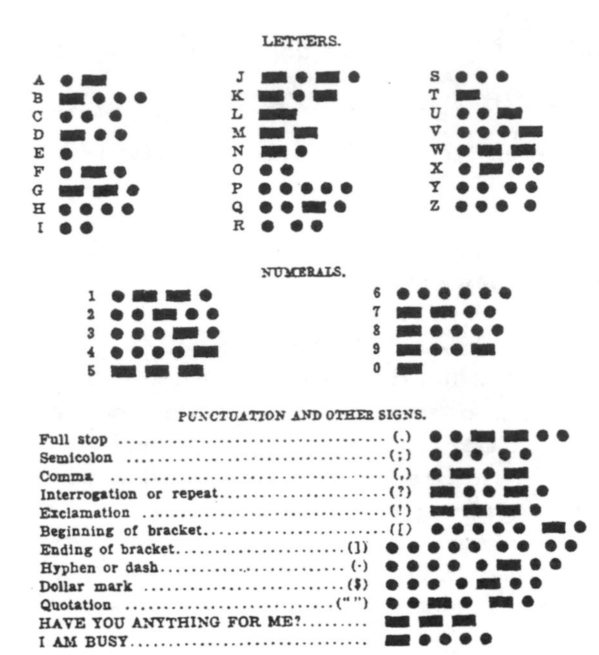

Fig. 8.5
Original Morse Code
(History of Telegraphy:2001-56)

currents seemed to be the only solution and that seemed to become more and more unreasonable with distance. Mr. Edward Davy's contribution to this matter was the electrical relay (Beauchamp, 2001:32). This device allowed the receiving current within the wire to activate an electromagnet that retransmitted the pulses exactly using a local battery source. This removed the necessity for large currents and formed a repeating method that supplied an isolated current at each relay point (fig. 8.6). In 1837, this patent was purchased by the Electric Telegraph Company.

In 1837, Mr. Morse applied for a patent to this invention and was refused by the English, who accurately stated that it had been patented already. The French, however, did issue a patent to Mr. Morse, who termed this device a *transmitting relay.* Because the transmitting relay was used by Morse and was an integral component of the telegraph system in general, he eliminated his competitors. Without it any hope of successful communication over an appreciable distance was impossible.

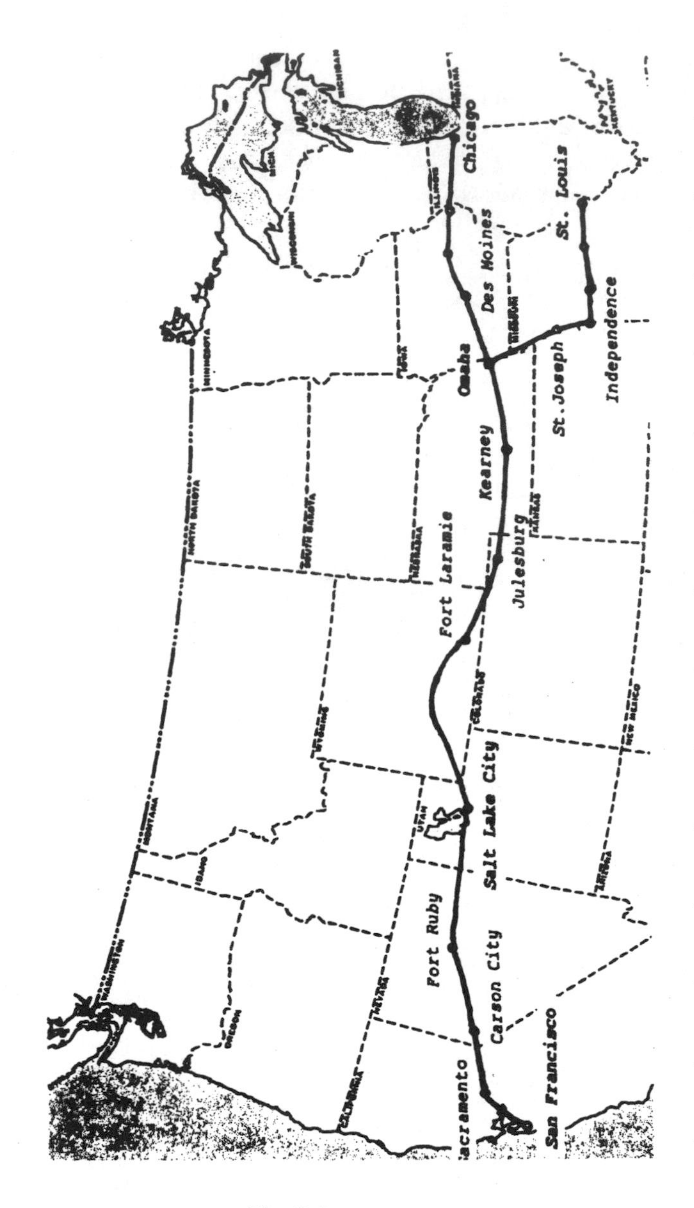

Fig. 8.6
Western and last link of the Telegraph 1861
The Telegraph: A History of Morse's Invention and its Predecessors in the United States copyright 1993 Lewis Coe by permission of McFarland & Company, Inc. Box 611, Jefferson NC 28640

9

The Acquisitions of Observations

Distant Territories

The transmission and receptive aspect of the telegraph was just as important as the fact that a standardized descriptive format was now available, This combination condensed the arrival and comprehension of an event from its occurrence to being almost instantaneous. However, one very important aspect of any event situation is the distance over which that event can be described. This is most important when there would be concern of that event reoccurring within high-risk probabilities, based upon physical conclusions that would promote its expectation.

The early telegraph was between Lancaster and Harrisburg, Pennsylvania, and at its completion in 1845 the public was not deeply concerned about the possibilities that it held for them. This, however, came to an immediate end during the late eighteenth and early nineteenth centuries. Two men who were deeply involved with the early telegraph traffic formed a company called Western Union, officially recognized on April 4, 1856 (Coe, 1993:86). This information is significant because without this company the transcontinental link between San Francisco and Washington would in most probability have not been constructed when it was.

By 1860, the farthest point that the telegraph extended was Omaha. At this time, the line was established along the route of the Pony Express, surveyed by Mr. Edward Creighton in 1860 (Coe, 1993:38); later the telegraph routing was moved whenever possible to be along that of the railway system.. After the remaining continental survey between Salt Lake City and California had been completed, Mr. Creighton met in California with Jeptha Wade, also of Western Union, in March of 1861 to discuss the remaining efforts for the last interconnection of the telegraph line between Salt Lake and San Francisco. Both men on April 12 returned to New York, and Mr. Creighton again left for Salt Lake to prepare for the final western construction. In the meantime, Mr. Sibley of Western Union had been successful in obtaining a grant from the U.S. Congress to establish the link by July 31, 1862. With funding and capital generated from Western Union stock, the eastbound construction crew left Sacramento on May 27, 1861, and the westbound crew set the first pole and left Fort Kearney on July 4, 1861 (fig. 8.6) under the overall supervision of Mr. Creighton.

The on-site crew management was listed as being under Mr. I. M. Hubbard working eastward from Carson City, representing the Pacific Telegraph Company, and the westward construction crew was under the on-site management of Mr. James Street, working under the Overland Telegraph Company. It is important to note that the telegraph construction line was done via both Pacific Telegraph and Overland Telegraph Company crews, although the effort was supervised, organized, and funded by the Western Union Company. The first transcontinental message was sent from San Francisco on October 24, 1861, at 7:40 P.M. to President Abraham Lincoln from Mr. Horace W. Carpentier, president of the Overland Telegraph Company. With the line in place that would effectively provide a means of communicating information, it was just a matter of time before the possibility that the information being sent from the West to the East contained meteorological data.

The Observation

If a chart of the atmosphere is to have any meaning, it must present observed events relative to a subject in such manner, as to be understandable and informative to the reader. The dictionary indicates that a chart can direct, perform a function of navigation, inform, and be an explanation. I believe that in certain ways our current meteorological analysis may have included the preceding, with two exceptions First, all the information regarding atmospheric compositions has been explained. That is the atom and how it relates to our breathable nature. The history of how the atom came to be defined and why the science of the atmosphere was relevant to certain peoples and not others have also been analyzed. How the atoms combine with temperature to form clouds and the result of their densities upon us have been reviewed and explored in wind, pressure, temperature, and our terrestrial magnetic field All in a neat progression to explain what is around us that can be seen, as well as the unseen. All except the chart in graphic form has been presented to adequately stimulate an understanding of the atomic structures that cover such a big part of our lives.

A chart, or picture if you will, is worth one thousand words in such a way that it combines the structures of molecular formations to present their density and radiative qualities and would therefore have significant meaning in the graphic representation of previous chapters. However, to provide a representation of any meteorological formation requires two things. The first is an event that is observable and warranting its record, and the second is the graphical representation of that event itself.

An observation of a meteorological event that is meaningful to our society as a whole did not start with the Weather Channel; it started in the 1800s with the military and then the telegraph. In 1812, James Tilton, then surgeon general of the U.S. Army, began an observation program that mandated all hospitals under his command keep weather records, which eventually became ninety-seven bases, directed by James P. Espy, who in 1842 became meteorologist to the U.S. government by an act of Congress. By 1849, Mr Joseph Henry, who was then secretary of the Smithsonian Institution, enlarged the program under his directorship to 150 reporting observation stations (Phyllis, 1993:16, 17). Because the original telegraph was installed in 1845, it was available in 1849 to Mr. Henry, who utilized this new communication tool to gather climate data, and with the cooperation of Western Union every telegraph office was instructed to start the day with some form of climate observation. As the telegraph increased in length so did observations, until at one point there were 500 reporters and some centralized point was required.

The reporting was presented in a way that focused upon not only the War Department but also navigation along the seacoast and Great Lakes. This was important enough for Congress on February 9, 1870, to pass by joint resolution a bill forming the National Weather Service, introduced by meteorologist Increase Lapham (Phyllis, 1993:20). In 1871 this was expanded by the War Department to include observations extending beyond the surface conditions relating to temperature, wind direction, and barometric pressure readings. The idea was presented to expand the surface data to include higher-altitude measurements. One of the stations that was important in this requirement was constructed and operational in Christmas of 1870, that station was the weather observatory located at the top of Mount Washington in the state of New Hampshire at forty-four degrees, sixteen minuets north latitude and seventy-one degrees, eighteen minutes

Fig. 9.1
Pikes Peak Weather Station
(*Weather Pioneers*: 1993-33)

west longitude, at an elevation of 6,288 feet. An additional high-altitude station was constructed for atmospheric high-altitude purposes also but located in the Rocky Mountains at the summit of Pikes Peak (fig. 9.1) at an elevation of 14,134 feet becoming the first and highest station in the world. It became operational on October 11, 1873 (Phyllis, 1993:27). The Pikes Peak weather observatory is now out of existence; it was brought down to the ground in 1963, and a modern visitor's center in now in its place. The Washington observatory, however, is still most active in the transmission and research of high-altitude weather data.

The primary observations came from individuals located throughout the United States. As early as 1817, the General Land Office in the Treasury Department under the direction of its commissioner, Josiah Meigs, issued a directive to all office locations instructing the registers to keep a chronicle of such occurrences as the temperature, winds, and weather that had to be annotated into their daily log. Because the land offices were located in the areas of Michigan, Ohio, Indiana, Missouri, Louisiana and related weather observations to the areas of thirteen degrees latitude and ten degrees longitude, some coverage of that territory was available and recorded through the observers in their journals (Fleming, 1990:17).

Observations by the year of 1831 were numerous in every type of weather phenomenon without concrete causation, so much so that James P. Espy, then director of the Joint Committee in Philadelphia and an opponent to institutionalized meteorological observations, (Fleming, 1990:25) presented a theory that was instrumental in the convection theory associated with rotation and convergent air inducement systems as the causation of rain. This was a significant direction concerning the motivational aspects of a storm. Mr. Espy's conclusion was indeed very close to actual storm creations in that he surmised that the engine of storm creation was thermal heat. He correctly explained that as moisture-laden atmosphere was heated and by that process rose to higher altitudes where the atmosphere was thinner, it would expand, cool, and condense into a cloud. One of the most important details of this process was the fact that as the rising air expanded, it would leave a center of lower pressure at the surface, causing an imbalance in atmospheric pressure differences between the surface and some distance from the low's center. This difference caused the atmosphere to rush inward, replacing the rising air within the established column. This, he concluded, was the reason for the wind. His linkage between the wind and the inequalities of pressure within the rising column and the surrounding outside terrestrial area was the reason the manifestation of components consisting of temperature, density, barometric pressure, and moisture distribution is a fundamental cornerstone of today's meteorological studies (Fleming, 1990:26).

The Joint Committee that was formed with Mr. Espy received in 1834 some fourteen journals from across the country. However fledgling this was at the time, it should be understood that many observers had no instrumentation, formal training, or standard forms with which to correlate their findings into a database sufficient to form climatic conclusions. Yet as can be seen in figure 9.2, represented by the map produced and circulated by the committee in 1837, Mr. Espy's theory on wind direction does indeed bear evidence that it is rotational and convergent about the center of a storm's low-pressure area. Rudimentary as it was, this was indeed one of the first weather maps in that the wind direction was plotted as a function of meteorological observation and not as a climatic representation. Although the participation of observations across the country at the time of the Joint Committee publication was indeed important, because funding was not available through either the U.S. government or Institutions and the line of their authority and reporting structure was ill defined, the Joint Committee was dissolved in 1838. Monies that were appropriated were issued and deposited to the Franklin Institute because they were issuing and

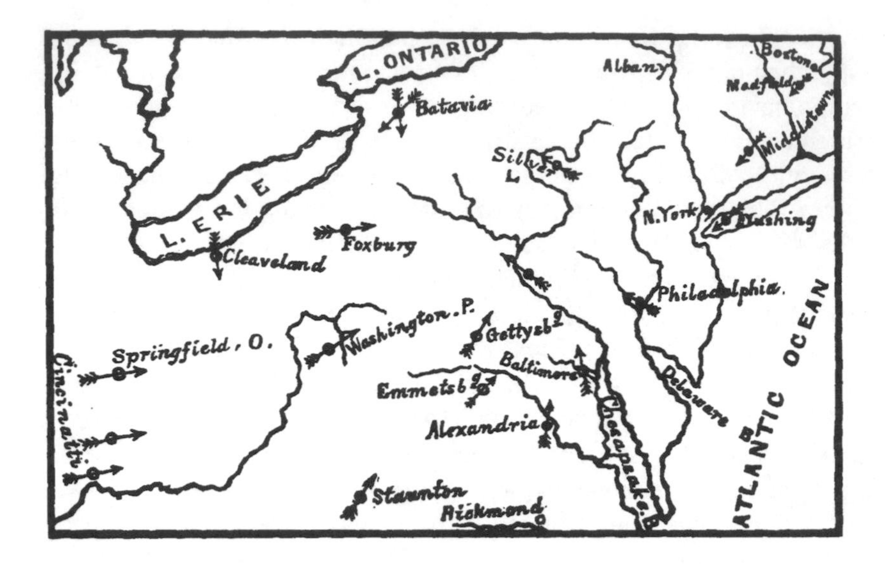

Fig. 9.2
Weather Map of Joint Committee, 1837
Converging Wind Pattern
(Meteorology in America, 1800-1870:1990-58)

paying the expenses for the distribution of directions, used as guidelines in instrumentation calibrations, instrument readings, and the methods used in the measurement of temperature, pressure, dew point, clearness of the sky, and whatever precipitation was occurring. Mr. Espy eventually became the meteorologist of the Joint Committee and in that capacity wrote an enhanced version of a meteorologists' guidebook titled *Hints to Observers in Meteorology* (1837), prompting observers to notice the storm's terrestrial path at different times of the year (seasons) as well as storm velocities and cloud shape. However, due to the lack of sufficient funding toward his contribution as a meteorologist, he turned toward the lecture circuit, presenting his theories on the possibility of creating rain through artificial means, upon the committee's demise in 1838.

Mr. Espy did not discontinue his endeavors for the advancement of meteorological studies and in time applied his perseverance through another government appointment. In August of 1842, he was directed by the surgeon general, Thomas Lawson, to report such observational activities at all military posts then in operation within the United States through a report submitted annually to the War Department. The first order that Lawson issued was undertaken quickly, and in December of that year Espy established an authority of observation control that was vast and extensive. Not only had Mr. Espy become a focus for report authority throughout all military posts, but his office claimed that within the military organization as a whole, since the military in general contained various force elements, an inclusion be considered that provided for naval installations, the ships of war, and coastal establishments applicable under navy command. This directive also included lighthouses and flotation lights as per authority of the secretary of the navy.

Additional influence to compile meteorological data was also rendered to Mr. Espy by the secretary of state with authority to establish a central Washington point of reception for observational meteorological data communicated from diplomatic sources with respect to U.S. embassies and staff, rendered by report to the Office of the Secretary of State with an eventual destination that of Mr. Espy's office. This was a foundation of extraordinary networking. Along with the increase of reporting prestige was an increase in army funding that enabled Mr. Espy to furnish twenty-four posts with a new instrument used to measure barometric pressure with readings taken to conform with the requirements of the Royal Society of London (Fleming, 1990:70).

The weather mapping of meteorological observations included within the report to the surgeon general's office was updated to include barometric pressure measurements along with wind direction. The map that compiled the observations included with Mr. Espy's report: *First Report on Meteorology to the Surgeon General of the United States Army,* dated 1843 and released in 1845, can be seen in figure 9.3, representing the additional lines of equal barometric pressure added to wind direction. The map suggests that in 1845 an analysis was graphically presented to depict barometric pressure lines, which today would be interpreted as isobars (Fleming, 1990:61).

The urgency for a centralized point at which the compilation of meteorological and climatological data was growing to a point beyond the detached elements of the Joint Committee, navy, Albany Institute, and army medical department's ability to coalesce. A new agency beyond the 1840s would be required to establish a national system to represent, control, and compile all the observational and historical data pertinent to atmospheric studies. A man would emerge who had already contributed heavily to the

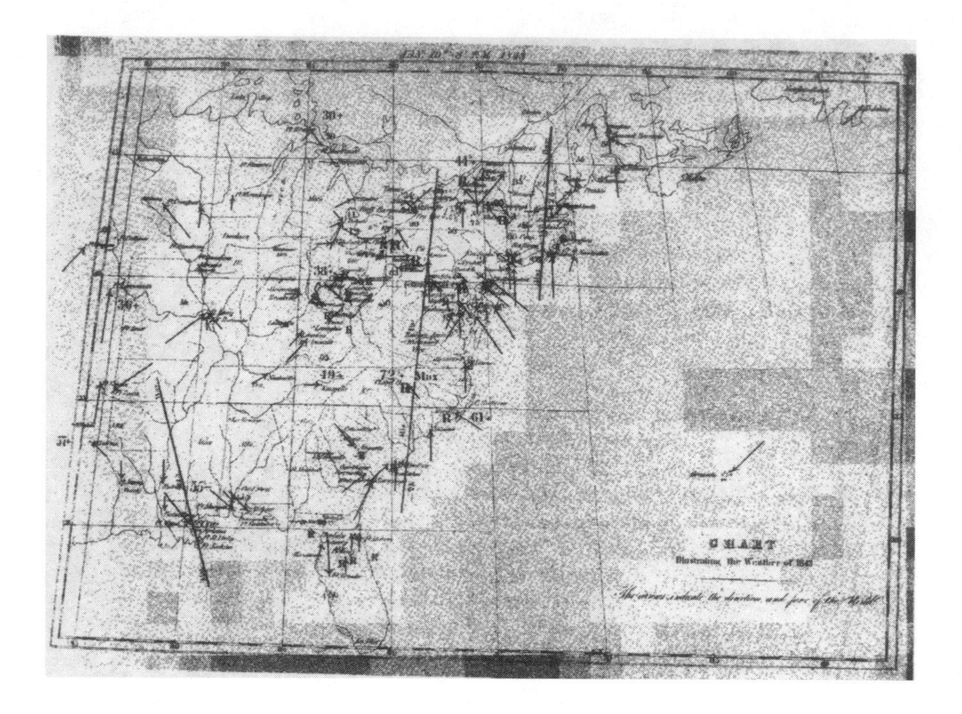

Fig 9.3
Mr. Espy's Meteorological Map as issued in 1845
(*Meteorology in America, 1800–1870*: 1990-71)

meteorological circle, although that contribution was more in the field of magnetism, and indeed his work in that field had been greatly significant in the role of communication within the meteorological discipline. Again Mr. Joseph Henry would ascend to another place within the meteorological system, and this time both his prominence and credit were irrefutable.

The Early Smithsonian and Others

It was at the Smithsonian that the observers' data was unified, it was this institution that charted the climate and weather of the new territories, it was this institution that pursued the theoretical creation of storms and became a national center for atmospheric research, and it was the Smithsonian that was a source for the calibration and distribution of instruments.

The Smithsonian provided a standardized form for observations that became affiliated guidelines and issued appropriate tables to facilitate in the correlation of meteorological findings.

Over six hundred observers located across the United States, Latin America, Mexico, and Canada kept journals to record meteorological data and sent to the Smithsonian Institute their reports via mail (Fleming, 1990:75). Under Mr. Henry, acting in the capacity of first secretary, the institute generated publications, advanced research, and provided lectures. The institute was also instrumental in other areas of national development, as was evident in its role of supplying meteorological data by an agreement in 1863 to the Department of Agriculture. These instruments were the consequence of a plan for instrument issuance by Mr. Henry that had been approved by the Smithsonian's board on December 15, 1847 (Monmonier, 1999:40). From 1850 to the 1870s, the Smithsonian Institution had observers who were segmented into three categories depending upon the instrumentation utilized. Category one contained those observers who did not have instrumentation; category two were observers who possessed a thermometer only, and the last category of observers were those who had a full set of instruments, including thermometer, barometer, rain gauge, and a psychrometer for the measurement of humidity (Monmonier, 1999:40). All observers were essentially located from the Ohio Valley to the Atlantic seaboard (fig. 9.4). Amateurs and scientists alike were joined by a common goal. That was to verify and historically provide accurate meteorological observations useful for the nation's advancement through science.

The Smithsonian Institute was founded in 1846 through funds allocated by the estate of the late James Smithsonian. The institute began with a building known as the Castle, erected and completed in 1855. In 1881 the building now known as Arts and Industries was erected, followed by the Zoological Park in 1891 and the Astrophysical Observatory, which was established in 1891 (Smithsonian Institute Archives/History Division, 1/31/2001). From a charting aspect, relative to atmospheric conditions presentable in a timely manor to the general public, the Smithsonian was only three years old when this milestone was reached through the creativity of Mr. Henry.

Because of the significance that the telegraph had for the transmission of real-time meteorological occurrences, and the subsequent variations in office geographic locale, the reports that were received by the Smithsonian were plotted on a large map that Mr. Henry had hung where visitors to the institute could view current data. The first plots consisted of the transmit-

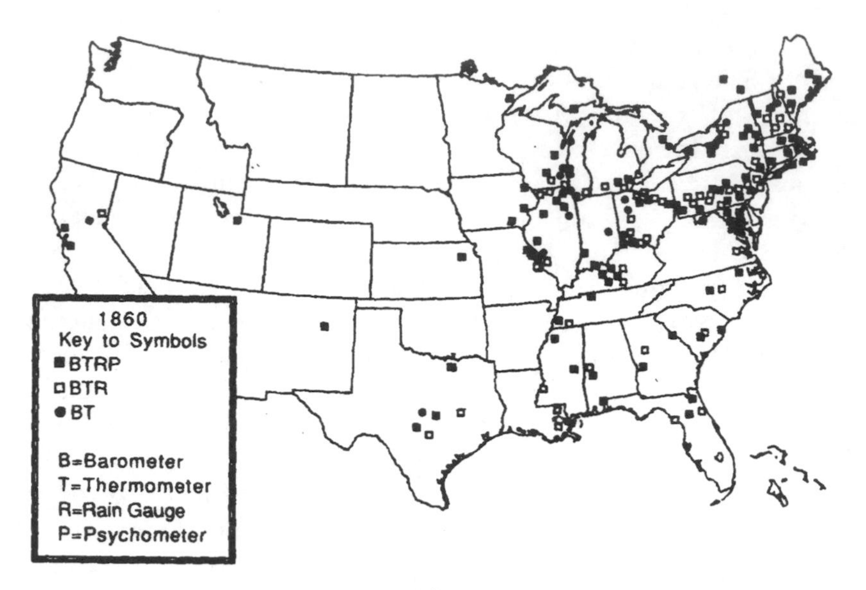

Fig. 9.4
Smithsonian Institute Observers (1860's)
(Meteorology in America, 1800-1870:1990-87)

ting stations' locations, which were identified with small colored disks embedded into the map with iron pins. The attached disks were of a color arrangement representing the current weather conditions at the time of the report, through the colors of black, gray, and white. These three colors would indicate rain (black), clouds (gray), and clear skies (white). Eventually the disks were upgraded to represent wind direction by means of an arrow that was inserted and pointed to one of eight holes that were placed around each disk's edge. This effort by Mr. Henry was indeed the first impetus that was of use in the development of a forecast. Using the wind and disk colors, some indication over time could be inferred as to the progression of conditions from west to east.

This attempt to actually forecast eventual conditions of weather-related meteorological progressions led to the Smithsonian Institute supplying this related information to the *Washington Evening Star* newspaper in 1857, and as a consequence of this cooperation the newspaper was one of the first to publish telegraphed weather reports (Monmonier, 1999:40). As important as the Smithsonian was to the emergence of the weather chart and its consequence for the forecast, this effort was not long-lasting. Congressional funding was paramount for the institute to continue in any role that was within the meteorological scope, and in 1865, in the fifteenth year of its operation, Mr. Henry partitioned Congress to relieve the observation burden that the Smithsonian Institute was currently undertaking and that a national weather service be created to assist in this endeavor. In addition, because the telegraph companies were effectively issuing reports from the field free of charge and on a voluntary basis, by 1868 this cooperation between the two organizations had largely began to dwindle. Mr. Henry tried to illicit funding through the Department of Agriculture because he felt very strongly that the farming and associated agrarian communities would benefit from meteorological reports. This, too, was not received in the light that Mr. Henry desired, and in 1872 the monthly report was discontinued by Commissioner Frederick Watts, along with the institute's franking privileges (Fleming, 1990:150). Eventually another organization would supersede the Smithsonian in chart and forecast development, annual budget allocations far exceeded that of the institute to a point where although the Smithsonian's allocations were never above $5,000, the newly formed U.S. Army Signal Office spent upward of $400,000 in 1874, and that was very early in the years of operation (1990:xix).

The Department of the Signal Office began on June 10, 1861, at Fort Monroe, Virginia, when Maj. Albert J. Myer reported for duty as the signal officer (Editors of the Army Times, 1961:11). The Signal Corps at that time did not have any complement of officers or men to provide the operation with any meaningful value. So it was that the first communication sent by Major Myer was a request for the detail of 2d Lt. Samuel T. Cushing from his position with the Second Infantry, in the capacity of signal duty, to be assigned to Fort Monroe for the purpose of exercising his signal experience until further officers may be trained in that capacity.

Major Myer was deeply interested in communication, and while at Graduate School in the Medical College at Buffalo, New York, he became most interested in the prospect that a form of communication for the deaf could be developed, which would be accurate in the conveyance of thoughts, wants, and desires. His graduate thesis embodied this concept in a proposal titled "A Sign Language for Deaf Mutes" (Editors of the Army Times, 1961:12), which essentially developed a methodology for the transmission of messages without previously utilized concepts. The intriguing concept was unique because it did not require the use of sound, touching, or the written word. This interest was very much alive and well in his military capacity with the army in 1854, where he was serving as an assistant surgeon in the Medical Department. In addition, the meteorological aspect of this pie is exposed by the link to the fact that as an assistant surgeon his additional capacity was to submit weather reports to the surgeon general from the post of Fort Davis, Texas. Myer's experience while on patrol with his men was extremely meaningful, in that he was able to observe the Indians demonstrating a communicative language using cloth and a rudimentary flag.

Within two years of observations, Myer had devised a visible communication method of his own, which utilized the alphabet and was simple enough to satisfy the intent for use by teenagers without advanced education skills. The instruments that would be required to transmit any alpha character message were essentially any two lamps, torches, or flags. The transmission was a semaphore in nature and would require a combination of three motions. The first visible daylight signal flag was essentially a composite of two colors and could consist of a white background with a red-centered smaller square or a red color background with a smaller white center square. This signal method with semaphore motion was adopted by the army in 1858 and used as a communication during the border conflict with the Apaches.

This was very successful, and with most successful operations comes advancement. Two years later, at the rank of major, Myer was ordered to report to Fort Monroe and appointed as the army's first signal officer. Although his request for Lieutenant Cushing was delayed for a time, the bombardment of Fort Sumter and the commencement of the Civil War made rapid execution of a communication sent that night to Maj. Gen. F. Butler, commanding the Department of Virginia for an additional detail of ten officers and three physically capable men (Editors of the Army Times, 1961:13). The first class (refer to fig. 9.5) of the Signal Corps was started with the requested ten officers, but also an additional twenty-seven men were sent to report under Major Myer's command.

In the fall of 1863, Major Myer was appointed as the chief signal officer with the accompanying rank of colonel; however, due to differences on how the Signal Corps was to be utilized from a manpower standing, he was in conflict very often with the then secretary of war, Edwin M. Stanton, with that consequence of being relieved and sent into exile at the Department of the Mississippi. It was this assignment that prompted Colonel Myer to leave the army. However, in 1867, due to a change in the govern-

Fig. 9.5
First Class of the United States Signal Corps.
(*History of the United States Signal Corps*: 1961-13)

ment position at the cabinet level, that being a new secretary of war who was Colonel Myer's friend General Grant, Myer was reinstated back to the Signal Corps with the rank of brigadier general (permanent rank 6-16-1880) retroactive back to March 15, 1865 (Fleming, 1990:155). It was in 1867 that the Signal Corps was at its lowest point, since the rest of his command with all but one lieutenant and two clerks had been mustered out due to the war's end. General Myer was not without recourse and with his extensive background in reconnaissance and signal reporting he looked to the unification of telegraphy in its use for meteorological purposes in the progression and subsequent tracking of storms. General Myer had experience in reporting weather occurrences in his prior military years and felt strongly that this particular role would be significant from a military standpoint. Through the usage of the telegraph, General Myer understood very well that the reporting, analysis, statistics, and relevant progression of such events would be as useful to the military just as his Corps experience was in the reporting of enemy movements during the war. Because the expansion of the telegraph networks linking not only the West Coast, interior, and East Coast, additional lines were now expanded to include business centers and points of commercial possibilities; this was an exceptional opportunity to also add the numerous military posts within these territories. His proposal included a presentation of maps that detailed his idea, which he presented to Congressman Paine, who in turn accepted the general's proposal and, with the partnership of Sen. Henry Wilson, presented the Joint Resolution to Congress on February 2, 1870, for its approval. The resolution was passed with no debate and signed into law by the general's friend President Grant on February 9, 1870.

Now that the weather bill was officially law, General Myer submitted a budget and staff requests for the Signal Corps to be established with its headquarters at Fort Whipple, Virginia. The first graduation of observer sergeants occurred on October 10, 1870 with a complement of twenty-five. All had progressed through a six-week basic training course that included meteorology, instrumentation, and their operation and repair, as well as the general maintenance of telegraph equipment used for their reporting. Reading material included the analysis of the Smithsonian maps at that time as well as observation methodologies that had been prepared and published by the Signal Office. At graduation, they were assigned to their respective military posts and equipped with instrumentation that included a rain gauge, thermometer, barometer, hygrometer, and anemometer for the measurement of wind velocities. Accompanying the weather observation

Fig. 9.6
Signal Corps Observer Network, 1860
(Meteorology in America, 1800-1870:1990-158)

sergeants was a set of manuals that they could use for reference. These included but were not limited to papers on fluid motion that could be found in the *Mathematical Monthly* editions, the manual of signals, Guyot's tables, *Winds of the Northern Hemisphere* by Coffin, *Climatology* by Blodget, and a *Barometer Manual* from the London Meteorological Office. The observations were ordered to be taken by Washington time and done in a synchronous manner three times daily, once at 8:00 A.M., 6:00 P.M., and midnight. The times of the observations were not chosen for any meteorological bearing but because at these times the telegraph lines were not occupied by the great increase of business and general traffic. In the beginning of this observation period, telegraphy operation was under the authority of Western Union and observation locations (fig. 9.6) were chosen by their proximity to the lines themselves; later Myer would expand that network considerably into uncharted territories, lighthouses, and mountains (Pikes Peak). As far as the telegraphic portion of the Signal Corps was concerned, by the end of 1878 5,000 miles of this communication format had been extended throughout both the Texas and North Dakota territories and, to the Signal Corps Weather Service's regret, General Myer died after succumbing to a brief illness, at which point in time General Hazen was placed in charge as the new commander. As a tribute and an honor to the accomplishments of General Myer, Fort Whipple was renamed Fort Myer.

The U.S. Weather Service that had begun from the inception of the Signal Corps was reestablished under the Department of Agriculture on July 1, 1891, then again placed under the Department of Commerce, where it currently resides today. The format of communication invented by Albert Myer is still usable as an approved methodology of communication, and weather observations as we know in this century can be directly linked to his most valued efforts and to the U.S. Signal Corps.

From 1871 to 1940 was a period of meteorological expansion regarding weather-related services. In 1874 the remaining observers (383) were transferred from the Smithsonian to the newly formed Signal Corps. By 1885, eleven years later, Atlantic storm warnings were being issued with the cooperation of the British Meteorological Service; as previously mentioned, the Weather Service was transferred to the Department of Agriculture in 1891. Observations within the Weather Service continued, and the first hurricane warning service was established in 1896. Contributions from ballooning in the first decade of 1900 was significant, and in 1910 the first of what were to become analytical weather forecasts began to be issued by the Weather Bureau on a weekly basis. Aircraft observations for the Weather Bureau were initiated about 1914, and this also was incorporated into the weather family and was affected with earnest as issuances were of great importance to the military due to the First World War. The military benefited greatly from the Weather Bureau's aviation reports, and the necessity for civilian information soon became apparent. Because of this need, in 1926 the Weather Bureau received responsibility to issue weather data to the civilian populace through the Air Commerce Act. Telegraphy was the primary mechanism of report communication, which was certainly understandable due to the prior affiliation with the Signal Service. However, with the first teletype installed in 1928, this means of communication abruptly ended in favor of the automatic printout (Shea, 1987:10).

The 1930s were significant for weather research with the installation of equipment termed *radiosondes,* which allow for parametric measurement of pressure, temperature, humidity, wind direction, and its velocity component. In 1940 the Department of Commerce was given the overall responsibility for the Weather Service through service transference out from under the Department of Agriculture. It was President Roosevelt's reasoning that this move would provide for greater government interagency cooperation and provide a greater benefit to the areas of aviation and other general commerce sectors.

The teletype was upgraded to accept facsimile transmission in 1948, which greatly improved the analysis of atmospheric, oceanic, and general data transmission. Additional upgrades to the teletype were affected in 1954 that enabled automatic sensory equipment to link directly with teletype transmissions began operation and were coordinated with early radar units that augmented the observational tracking and progression of hurricane situations occurring along coastal regions. As weather observations were required to provide a greater amount of synoptic analysis, larger responsibilities were placed on the technological sectors to provide larger-scale enhancements. This was not necessarily relegated to faster data transfer between observation and points of analysis but, rather, a greater degree of topographical observation. Generally this means a greater altitude must be achieved to perceive finer synoptic patternation.

Atmospheric observations up to this period were merely stepping-stones to the eventual application of remote operations. Accomplishments would be undertaken that would provide optical surveillance and parametric sensor display from an Earth orbit perspective. This technology increased analysis capabilities and added accuracy in projections that dealt with the inherent characteristics of atmospheric dynamics. Also, the optical and spectral sensor range over a greater synoptic scaling enhanced the possibility of capturing meteorological occurrences on a real-time basis. In addition, prior to the new age of remote sensing, atmospheric observations, available in any reasonable format, were basically localized to the immediate area of the sensor array. Sensor range regarding altitude limitations was restricted to aircraft locale (several hundred miles), and this restriction would apply to balloon observations as well. In the case of balloons, they would destruct at altitude and jettison the equipment by either accident or preplanned drop causing a specific retrieval criterion to be employed. Controllable technology from greater distances could only take one direction if any meaningful and accurate sensor imagery for synoptic scale analysis was to be addressed. That direction was up.

With the unification of new lighter-weight equipment and transmission and reception capabilities developed around the late fifties and early sixties came the advancement of orbital capacities. In 1961, through the cooperation with both military, NASA, and private sectors, an undertaking commenced that would prove this concept a reality. The TIROS (Television InfraRed Observation Satellite) vehicle (fig. 9.7) was conceived to provide meteorological analysis from an orbital platform with a purpose to enhance the synoptic problem through observation means (Shea, 1987:12). In 1963, the program successfully orbited *TIROS 8* and demonstrated the capability of that fundamental link. It was the solution that completed the unification between controllable Earth orbit technology and optics through transmitting the first television picture of our terrestrial atmosphere on a continuous basis. This feat was accomplished through the establishment of global downlink receiving stations located along the vehicle's orbital trajectory. Total, accurate, and real-time imagery of our terrestrial atmosphere and the general planetary surface had become a reality from a distance that encompassed in one picture many thousands of words. This was the optimal observation platform that would revolutionize meteorological occurrences in that they could be tracked from inception to decay.

With these events, the Weather Bureau's Office was eventually enrolled under the Institute for Atmospheric Sciences, and in 1967 this name and organization changed again to become one of eleven laboratories managed and combined under the designation of the Environmental Science Services Administration (ESSA).

Science and engineering progressed during the late sixties on placing platforms that could remain in a stationary position above and synchronous with the Earth's rotation. This orbital placement was called geostationary because the vehicles' rotation velocity matched the Earth's so it would remain above a fixed planetary surface point. However, to accomplish this point in space relative to the Earth required that a very significant orbital altitude be achieved. With the advancement of greater rocket thrust packages, orbital distances required to achieve this point in space were possible. Considering these advances in the atmospheric sciences along with many other responsibilities, President Nixon in 1970 established controlling authority that would among other things manage and develop the Weather Service. That program and department was called the National Oceanic and Atmospheric Administration (NOAA).

This organization would unify not only the weather sciences but many of the government's other agencies that heretofore had been responsible for environmental and other national activities related to the geoplanetary environment. NOAA would coalesce these agencies for the

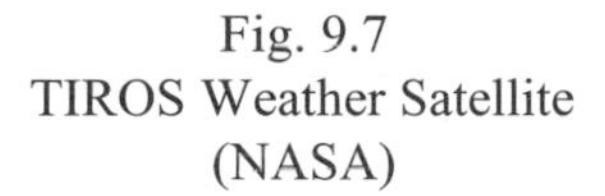

Fig. 9.7
TIROS Weather Satellite
(NASA)

purpose of establishing a systematic and rational approach to monitor, protect, understand, and develop the total geoplanetary environmental Earth system on a national level.

In 1974 and 1975, satellites with geosynchronous capability termed *Synchronous Meteorological Satellites (SMS)* began operation. Eventually, with these successes, the applicable term that we understand today as *Geostationary Operational Environmental Satellite (GOES)* is still in use (Shea, 1987:14). The current orbital distance for these space-based platforms is 22,000 miles, and they were positioned at a fixed point above the Equator to watch atmospheric activity such as hurricanes and tropical storms with continuous monitoring capability. There are currently satellites in a lower orbit above the planet at 522 miles; the intent for these vehicles is to observe atmospheric parameters associated with temperature, cloud cover, and the global environment that would be particular to surface conditions and changes.

In 1979, by presidential directive, a new orbital platform was designed to provide low-orbit controllable technology that observed through sensor arrays the terrestrial surface. That program was termed *LANDSAT,* which is under the management of a commercial organization called Earth Observation Satellite Company (EOSAT) and became operational on October 18, 1985.

10

A New Ability to Perceive

Sensor Considerations

In Chapter 7, there was an exposition relative to the properties that water vapor, nitrogen, and oxygen exhibit when subjected to electromagnetic radiation. Discussions about the sun's radiative bombardment and the absorption that clouds have as manifested by the shadowing effect beneath a cloud were undertaken. This is an important observation because it leads us to consider that shadows are a visible relationship to us optically of an atomic occurrence at the molecular level. Taking the optical effect one step further, consider that we only observe a small part of the electromagnetic spectrum and as thus would only be aware of additional manifestations through other tactile senses such as heat and touch (moisture) but not aware optically. What if a false image of either the atomic or molecular responses to an electromagnetic stimulus could be produced in such a way as to visually see what can't be seen normally? Of course for this reasoning to be possible, some method would be necessary to, in effect, measure a variance from some standard reference in the electromagnetic spectrum. It can be remembered that atomic structures will react when subject to certain wavelengths of energy through orbital changes of the valance electrons. If an electron absorbs energy it will change its relative distance to the nucleus and emit some photonic quantum. In Chapter 2, I discussed the Bohr atom in such a way as to illustrate this electron property. The electron will emit a frequency that is the difference between an attained orbit minus the prior orbit's energy level state. Since this is true, frequencies within the electromagnetic spectrum could quite possibly identify electrons in orbital exchanges, and if these frequencies are measurable, then some form of imagery could be designed to depict this property and, taken further, be displayable within our optical spectrum for analysis and observation.

Obviously a relationship exists that will provide some reference to spectral considerations if the quantums of energy expected are proportional to the orbital levels of the atomic and molecular structures in question. Since we have identified the atmosphere as to its constituents and their arrangement, then it is most reasonable to predict the frequencies of radiative quantums that would be produced at certain electromagnetic bombardment levels.

To be knowledgeable in satellite meteorology and a vehicle's relationship to observation quanta within and about the planetary atmosphere, one must have a basic concept of the various ways that any emission or absorption is receivable. The science that studies either emission or absorption by gases is called spectroscopy (Kidder and Haar, 1995:64) and is the primary science that is utilized when discussions of a radiative interaction with atomic structures are illustrated. There are four basic ways that radiation can be altered as it passes through a material (in this case the material is the atmosphere). First, radiation from a transmitting source can be absorbed by the material that is conducting the focused energy; second, the conducting material can emit radiation by its interaction with the focused energy. Third, the focused energy can be redirected and scattered out of its trajectory into other directions when it interacts with the target material; and fourth, the energy that is created through emissions from other materials can be scattered or directed into the focused path. These four possibilities should be considered when focused energy passes through a material.

Radiation or, in this case, radiation energy that is removed from the focused path or in some way is redirected out of the path is called depletion, and the radiation or emitted energy that is received into the path of focused energy, adding to its total radiation, is called source terms (Kidder and Haar, 1995:58). The law that covers any decrease of focused energy as it passes through a medium is Beer's Law, which states that the rate of decrease in the intensity of radiation passing through a medium is proportional to the intensity of the radiation. We can understand this activity in terms of a conducting medium generating radiation that will pass through the planetary atmosphere to a vehicle's sensor.

In consideration of radiation that is ascending up to a sensor from the atmospheric medium, if the medium is transparent then there would be no expected change to the ascending radiation energy as a differentiation from its source; however, if the atmospheric medium is not transparent, that is to say, there is some matter that is affected by the ascending radiation, then a variable will exist between the source and the received sensor

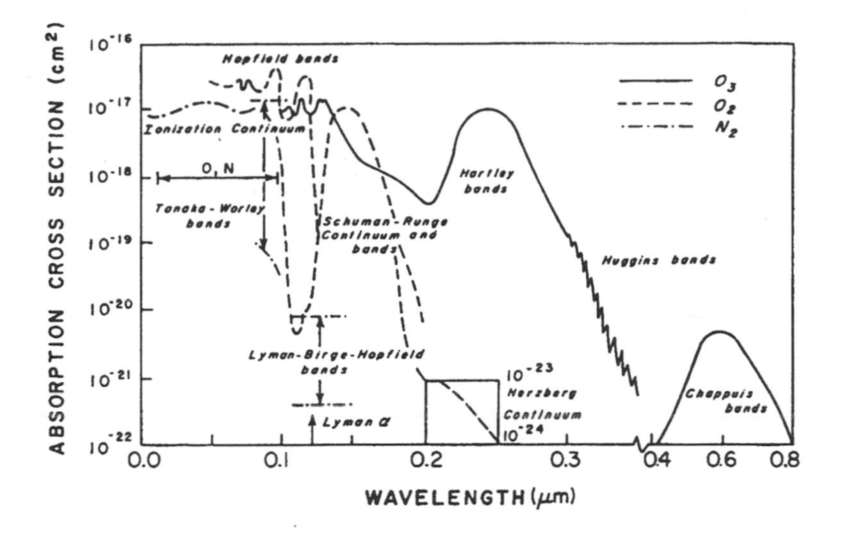

Fig. 10.1
Molecular Atmospheric Gas Absorption in UV and Visaable Spectrum
(Satellite Meteorology/An Introduction:1995-65)
* relative importance, not necessarily quantitative

value. That variable will manifest itself through a temperature and radiance change in the ascending beam. It is important to note that radiation will affect the atmospheric medium in several ways. Molecular and atomic structures will undergo a change in their vibratory states or a change in the structures themselves.

It is because of the ionization of molecular structures that electrons can be removed from their orbits at wavelengths of the ultraviolet and shorter and can be applicable to most of the ultraviolet wavelengths being absorbed by the O_3, O_2, and N_2 molecular structures within our planetary atmosphere (fig. 10.1).

A change in the motion state of a molecule is a result of radiation with regard to vibrational states (figs. 4.7 and 4.8) and, will occur in wavelengths of the infrared. The primary molecules affected by a change in motion when infrared radiation is absorbed are carbon dioxide and water vapor. Note, since radiation is a form of polarized energy, if a molecule does not possess an electric dipole moment, then it will not interact. This will go a long way in explaining why infrared radiation has no effect on the molecules of N_2 and O_2 and, they behave transparently to it.

Behaviors of molecular activity such as structure rearrangement through dissociation and motion effects manifested by vibration changes can be viewed from an orbital platform and through those observations explain many circumstances that arise in our atmosphere and become charitable through that imagery. Due to the instrumentation on board orbital meteorological platforms, measurements of the unseen such as mentioned earlier are accomplished from great distances and presented in formats transmitted to ground stations that will further process the data stream into pictures and line images familiar to us.

The sensors themselves bear scrutiny because information as to how they receive invisible radiation from our earth in an upward manner and how that reception is translated into a transmittal energy format will add to the already expanding background of this book. It is important to understand what is received and the process that that data undergoes, which will enable us to connect the links between the observation itself and the results of that interpretation into a readable and understandable format, which for us results in the eventual charting of it.

The Blackbody

A perfect radiator is a material or body that absorbs all the radiation presented (incident) at every wavelength that exists and emits all the radiation it has absorbed at all wavelengths in existence. In addition, because all the radiation incident upon the body was absorbed, then no radiation was reflected. All this occurs at a specific temperature, and therefore the object is considered to be defined as a blackbody (1991:31). Do not be misled by the use of the color identification of black; many colors can be identified by the blackbody law and are not black in color. In this case, both the sun and our Earth are considered to be defined by these laws of radiation. The sun approximates a blackbody at the temperature of 6,000 degrees Kelvin and closely matches the Earth's received radiation or irradiance (wavelength of .5 um or the green region of the visible) at the top of the atmosphere (fig. 10.2). As far as the Earth is concerned, we may call our planet a blackbody because it also emits radiation. Although it is not as hot thermally as the sun, the emitted radiation will be at a higher number and longer wavelength, producing a correspondingly lower frequency in hertz (fig. 2.15) within the electromagnetic spectrum. This is identified as the region of the invisible, approximated through the illustration of figure 10.3 and governed by Wien's Displacement Law (1991:32). This law in physics states that the intensity (projected power, i.e., strength) of the radiation emitted is inversely proportional to the temperature (thermal measurement) of the object. The sun's temperature is vastly hotter (greater ability to project electromagnetic energy) than our Earth's, so its radiation wavelength number would be inversely proportional (lower and shorter by the same magnitude as in figure 10.2, resulting in a higher frequency in hertz) in a relationship its temperature.

Lower numbers, shorter wavelengths of light, i.e., thermal properties for intensity, require greater temperatures to achieve a greater output (fig 10.2). The larger the wavelength number to be projected, the lower the temperature required and therefore, lower intensity to achieve emission using a blackbody of 300 kilograms as the Earth (fig. 10.3). Put another way, it is easier for a cooler blackbody temperature to radiate higher wavelength numbers, and longer wavelengths of light, if the intensity derived from temperature is not available. Therefore, because the Earth is a blackbody, however cooler, in the sense that it emits wavelengths of light, its peak electromagnetic spectral range would be expected to be a higher wavelength number with a longer wavelength and within a lower frequency region (fig. 10.3).

In review of this important understanding, it is because the sun is a vast heat source (atomic fusion) that the surface temperatures stimulate the intensity power (ability to project outward) and therefore would radiate its greatest energy output in areas of the visible or lower number and a shorter wavelength (.4 micrometer) on the wavelength scale and at a higher frequency (fig. 10.2). In contrast, because the Earth is not as thermally hot as the sun, its ability to project radiation outward (projected power capability) is confined to less available energy and will radiate in higher wavelength numbers, with longer wavelengths of light (approximately ten micrometers) such as the infrared or invisible with a lower frequency range (fig. 10.3). It is because of this physical law that satellites are able to measure the Earth's received solar radiation and its radiative emissions back to space.

This relationship to temperature and the layers in molecular volume that the atmosphere presents then becomes fundamental in an ability to remotely observe meteorological events, using the electromagnetic spec-

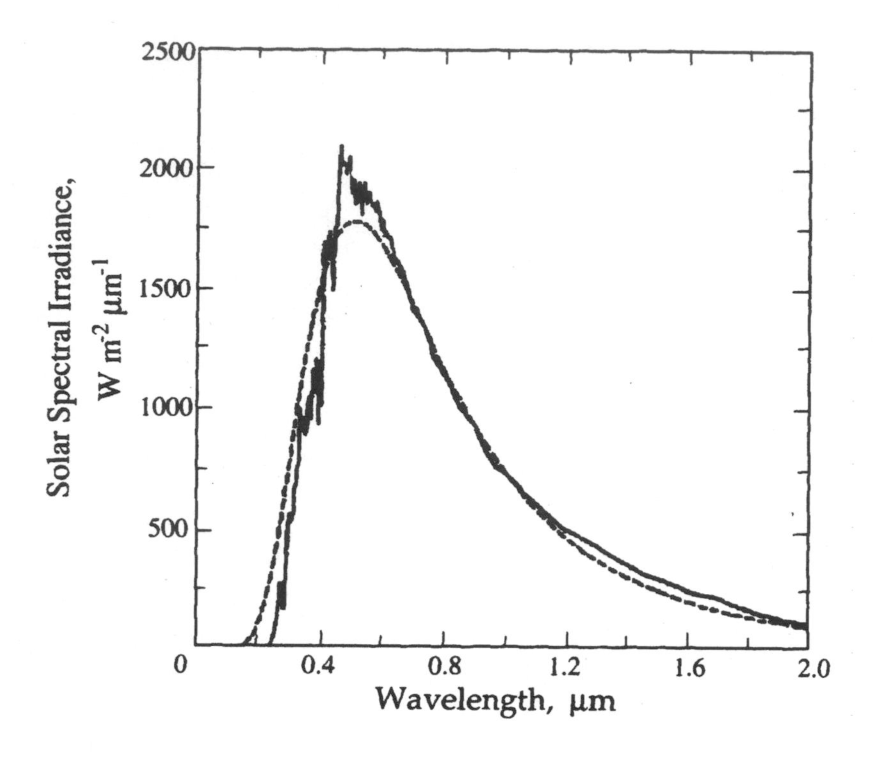

Fig. 10.2
The Intensity Radiation, Earth vs Sun at
the Top of the Earth's Atmosphere
(Atmospheric Chemistry and Physics:1998-26)

trum. When we discuss the spectrum of electromagnetic radiation in terms of units received and units emitted back to space by our earth, if we say that the totality of solar radiation being received by the Earth is 100, then it would be understandable that not all of the 100 units would be emitted back because there will be specific factors that will influence that emission. Six units are emitted back to space by the backscattering of clouds, then twenty units are emitted back to space by reflected radiation from clouds, and four units are emitted back to space from the surface of this planet (1992:94). This scattered or reflected solar radiation is termed *planetary albedo.*

Additional radiation is emitted back to space is separate longwave ra-

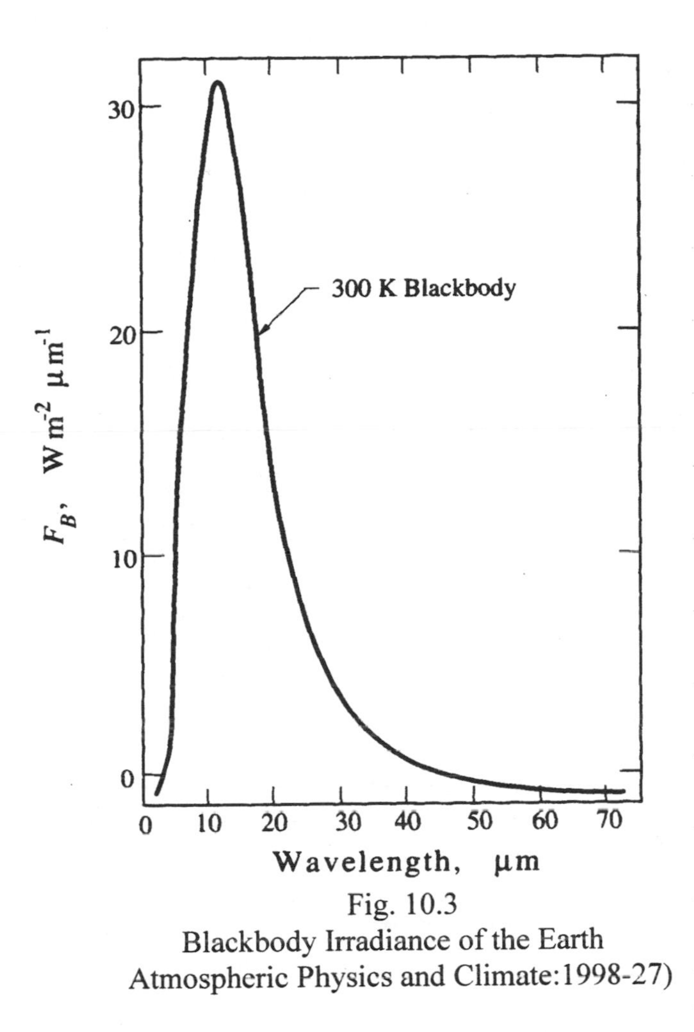

Fig. 10.3
Blackbody Irradiance of the Earth
Atmospheric Physics and Climate:1998-27)

diation. Part of emissions back to space are six units reflected, thirty-eight units from water and carbon dioxide, and twenty-six units from clouds (1992:94).

The remainder of received solar radiation that is absorbed from the atmospheric constituents of water vapor at sixteen units and three units is absorbed by clouds, and fifty-one units of solar radiation are absorbed by the planetary surface. Absorption from matter comprising our atmospheric components and its relative amounts, 1 viewed as 100 percent and 0 viewed as 0 percent, are illustrated in fig. 10.4, where the totality of absorption for all components is the lowest band and component matter is broken out in succeeding higher bands. "Vibrational transitions are the basis for temperature soundings" (71).

Electromagnetic Dissemination

Relative to the emission of radiation, not all of the electromagnetic radiation that is emitted will go straight back to the observing vehicle, but it will instead propagate out in all directions. The term that encompasses this physical property is *scattering,* and it can be divided into Negligible, Rayleigh, Mie 1, Mie 2, and Geometric Optics. Scattering is a function of several things. One is the size of the sphere (fig. 10.5) or particle, two is the particular shape that the particle has and is dependent on the specific wavelength that the particle is subject to, the viewing angle or the geometry of the received radiation with respect to the particle surface, and the third is the index of refraction or the ratio of the speed at which the radiation travels within the particle to the speed of light as referenced to a vacuum. At sea level, the index of refraction for air is 1.003, or about the same as that which would exist in a vacuum; however, in the case of a strong vertical gradient in discussions of atmospheric density some bending of the electromagnetic energy may result in a slight offset from the surface to the vehicle sensor in a positioning relationship (Kidder and Haar, 1995:48). For the various forms illustrated (fig. 10.5), Rayleigh Scattering is not in effect for particle sizes less than 10^{-3} and is only available in the spectral region of shorter wavelengths such as visible and ultraviolet. This explains a great deal about why we see the color blue in our atmosphere, and it may be stated that the smaller the particle, the more scattered the shorter wavelengths of the electromagnetic spectrum will be (1992:102).

Mie Scattering is a property where water drops such as rain and aerosols such as gas and dust will react with radiation in a greater way to create a diffused reflection. This is the reason that we have rainbows, halos, and coronas. For sizes greater than 50, where the sphere is very large in comparison to the wavelength upon it, the term *Geometric Optics* applies. In this manner the radiation is not refracted or reflected inside the sphere as in lower sizes but is reflected off the surface of the scatterer, and refraction

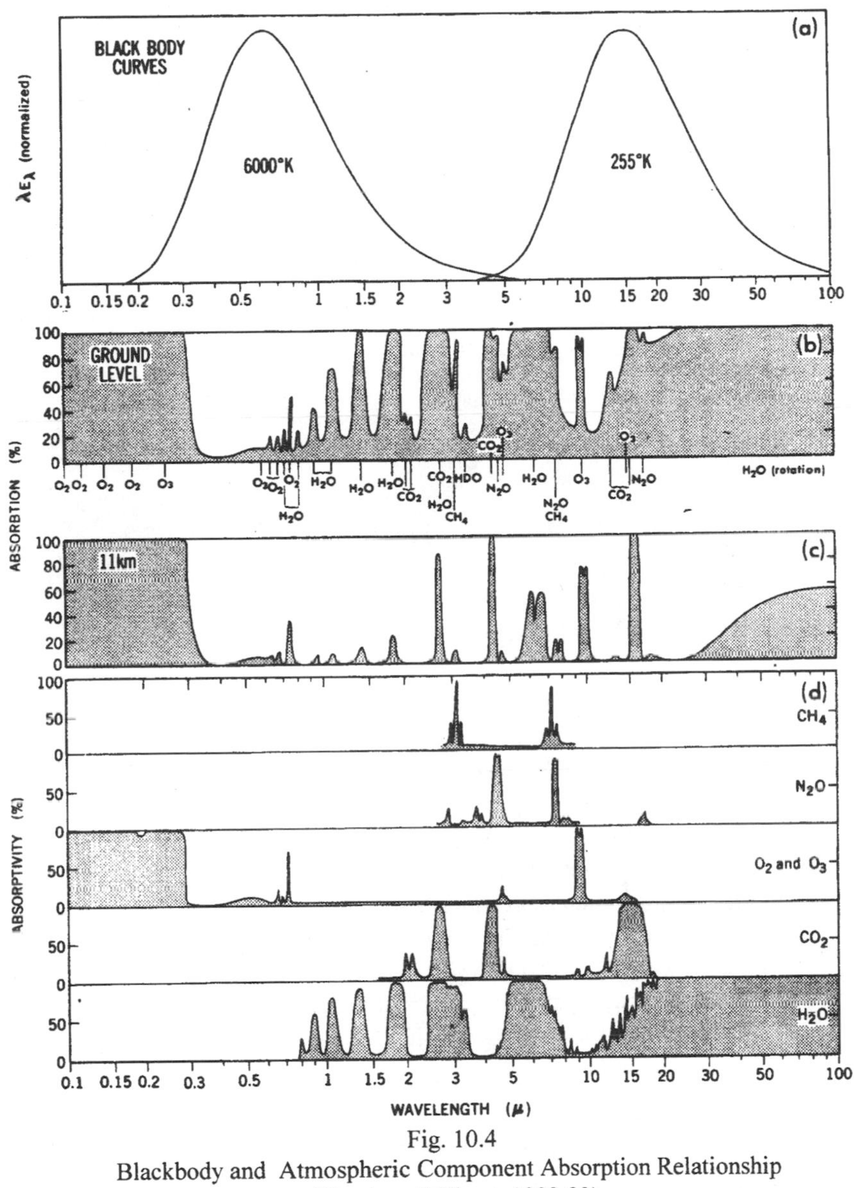

Fig. 10.4
Blackbody and Atmospheric Component Absorption Relationship
(Physics of Climate1992:93)

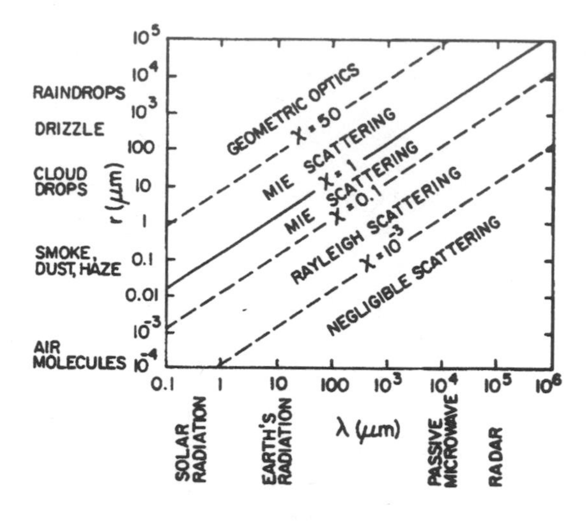

Fig. 10.5

Scattering Regimes for Particle Size
(Satellite Meteorology:1995-73)

will occur if the particle is a nonspheroid. I have used the word *refraction* several times, and for those readers not familiar with this word a bit of clarification will be in order. Light rays will bend as they pass from one transparent medium to another; in the situation between air and that of water, light rays will bend as they penetrate the boundary surface. This bending of light rays is called refraction, and, in the case of water, if you look at an object on an angle as it lies beneath the surface, a fish, for example, the position of that fish will not be where you think it is from your location looking at it over the side of the boat. The bending of the light ray is due to the fact that it travels faster in air by a fractional amount than through ice or water (see the index of refraction). This bending effect will only occur if the viewing angle to the object is different from ninety degrees or less than the perpendicular.

Reflection is different from refraction. In the situation of reflection, as the rays of electromagnetic energy meet the interface between two mediums, i.e., air and cloud surface, a portion of the bombarding radiation is thrown back at an angle that is equal to the angle measured from the perpendicular to the striking ray or angle of incidence. This is called the law of refraction (1991:41). It because of this physical property that a wide variety of optical phenomena such as rainbows and halos will occur from the hydrometeoric shapes (Kidder and Haar, 1995:74). Since the longer wavelengths of the electromagnetic spectrum will interact because of larger particle sizes, the spectrum of the infrared can be used for their identification and detection.

Clouds are vastly important to high-altitude observation, and because of the various particle sizes, from 10 um to 1 millimeter, they constitute a large interaction with the scattering of visible radiation. Also, because water absorbs spectral wavelengths outside the visible light portion, a great deal of visible light scattering will take place and absorption of the infrared will be almost 100 percent. In addition, they will conduct as blackbodies. In relative terms of clouds, the nonwater types are almost transparent to the short wavelength of microwave. However, a rain-containing cloud is not transparent and forms the foundation for using microwave radiation in the detection of clouds that possess large quantities of rain-size particles At higher altitudes where the formations of water vapor are in fact ice crystals, absorption is less than if the formations were of the more significant water containment variety. Also, the fact that high cirrus clouds would be thinner than a water cloud is also contributory to a higher transmittance factor because, they will have fewer particles per square unit of volume than the larger, more extensive rain-containing cloud formations. All these factors are very relevant to the scattering of incident solar radiation upon the planetary and atmospheric surfaces. Both blackbody and scattering relationships to the electromagnetic spectrum are vital when considerations of their retrieval from the refractive and reflective characteristics that or biting platforms would receive are undertaken. The interpretation of this radiation and, in some ways, how the radiation is received and identified due to the imaged characteristics themselves is important to the eventual analysis and imagery that provides for their identification and the eventual charting from that observation data.

Spectral Detection from Space

Orbiting vehicles that have the sole purpose the detection and onboard conversion of atmospheric data are an important link in the synoptic picture of planetary conditions. Observations from these stations have indeed come a long way from the first concepts that Mr. Henry envisioned in the early 1800s.

Energy measurements have a variety of meanings, from temperature ratios to vapor densities. All are relative to the acquisition of meteorological data. All are important and necessary in any understanding of atmospheric properties and the forecast of their behavior. In the case of acquisition vehicles tracking across the surface from an orbital standpoint, the Earth passes beneath the vehicle and the vehicle is orientated in a configuration that allows for the scanning instrument to be mounted underneath it to acquire data in a fashion that is perpendicular to the ground track. If the orbit is circular in nature, then some capability exists that offers an amount of predetermination in subsequent calculations relative to additional retrieval possibilities, for some low Earth platforms are configured in such a way as to provide for an overlapping of data on not only previous but also successive passes (Kidder and Haar, 1995:42). Data, that is, of the radiation that is viewed by the sensor array, can be said to be retrieved through mirrors. The first surface that is impinged with spectral Earth energies is elliptical with respect to the axis of the primary mirror and rotates at 360 revolutions per minute. This is not by chance; the scan rotation is calculated with considerations of vehicle velocity and, this produces one scan every 1.1 kilometer between subpoints (TIROS N series), the subpoint being an intersection between the perpendicular scan area and its midintersection point on the ground track (subpoint). This method can be found in use for NOAA platforms as well and is termed *AVHRR,* or Very High Resolution Radiometer. The sample of electromagnetic scan is focused upon a primary mirror as stated and is then redirected to a focal mirror (secondary mirror) that again directs the energy into a small center point hole within the primary mirror for dissemination by further optical instrumentation. The purpose for spectral deflection is to produce a refinement of spectral quality in such a way as to isolate the wavelengths for individual transmissions on separate radio channels (fig. 10.6). The electromagnetic energy that strikes the sensor is converted to an analog voltage that is proportional to the radiance received from the viewing area. That voltage is digitized into a data bit word that totals ten bits or a representation of between 0 and 1,023 possible values. (The electrical engineering that describes this conversion is not part of the scope of this book.) Each ten-bit word is considered to be one pixel (photographic point) of image and so transmitted to the ground receiving station.

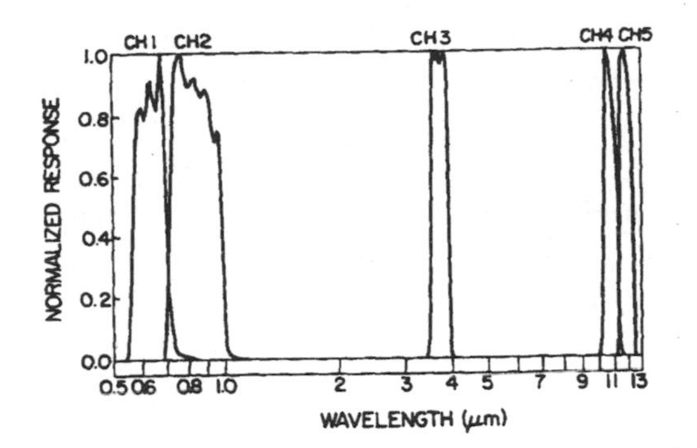

Fig. 10.6
AVHRR Spectral Responce
(Satellite Meteorology:1995-94)

Calibration is required for accurate and precise energy measurements, and therefore a form of reference is required. Some of the reference values are programmed into the electronics prior to vehicle launch, while some of the reference values are extracted on board the platform through quartz-iodide lamps that when energized will emit a specific radiance value. Since the platform contains twelve such lamps, the combination of multiple selections will provide spectral references that can be utilized for comparison against the viewing sample. Other channels are referenced against a hot or cold surface, and reference values are extracted during the viewing mirror's rotation, as it views the cold of space and instrument housing. Radiance values are established providing references from extrapolations, using the blackbody relationship determined by the Planck equations. Picture images that are transmitted to Earth receiving stations that comply with the TIROS-N series are in two formats. The first to discuss is the high-resolution picture transmission (HRPT), and the second is termed *automatic picture transmission (APT)* (Kidder and Haar, 1995:97). Both data information formats are conveyed through separate transmitters, each with a power output of approximately five watts. Unlike the high-resolution transmission, the second form of data format is in a variable voltage stream and not in a digitized word format. This voltage format is called automatic picture transmission, or APT.

The infrared scanning instrumentation utilized for radiance retrieval is similar to the AVHRR unit with differences; the acronym for it is *HIRS* (High Infrared Radiation Sounder). This device is flown on the Nimbus program and provides for expanded telemetry channels with a scan area of 42 kilometers rather than the 1.1 kilometer on AVHRR. A basic difference is in how the two packages of instrumentation are employed. The AVHRR unit profiles the horizontal altering of the atmosphere, and the HIRS unit is employed to sound or profile the vertical structure of the atmosphere. Varieties of profiling capability are possible. Moisture in water vapor is sensed, ozone is profiled, and so is cloud detection. Mirror rotation on these devices is incorporated as well, except this mirror rotates in steps producing an oval area that resembles a rectangular patchwork consisting of oval shapes throughout the scan area of 100 steps. All twenty channels are enabled at each step scan. Unlike the AVHRR unit, which calibrates at each scan line, the HIRS unit self-calibrates every 256 scans. The reference to cold space is also employed. As with the AVHRR, a digitization of the voltages determined is done prior to Earth transmission.

NOAA (National Oceanic and Atmospheric Administration) uses a Microwave Sounding Unit (MSU) on board their platforms. This instrument provides meteorological information with regard to temperature profiles in and out of the presence of clouds. The detection of microwave lengths in this sounding process is at best more difficult than at shorter wavelength energies. Antenna sizes are a ratio to the wavelength of frequency, so the vast difference in size between infrared at ten micrometers and the shorter wavelength of microwave at one centimeter constitutes 1/1,000 of a meter. The voltage difference measured once the microwave energy has been received on board is mixed with a known radio frequency. The output from that mixing is a ratio of proportions between the brightness received and an internal brightness reference. Because microwave receivers have a greater sensitivity than that of infrared reception and Earth radiance is less for microwave than infrared, more of the Microwave Sounding Units are being utilized (Kidder and Haar, 1995:105).

The application for atmospheric ultraviolet sensory retrieval has a unique place in our atmospheric understanding. Because the solar spectrum of electromagnetic terrestrial radiation has within it this most harmful region, the oxygen atoms that form the O_3 molecule of ozone, and O_2, make observations of this radiation most valuable. The terms *photodissociation* and *photoionization* apply greatly here and enhance the previous discussion on them. In chapter 4 some attention was given to the atom of oxygen and the molecular structures that it would form within the atmospheric envelope.

The creation of another molecular byproduct or particle removal that creates an ionic state with a distinct polarity that is an attraction or repellable to other unbalanced atomic configurations is a basis to understanding the ultraviolet wavelength. Harmful manifestations to living tissue are possible as the subject atoms within the cellular structure vibrate through the absorption process and themselves became dissociated through a process that incorporates a difference in energy levels from a specific lower level to that of a higher level of energy than would be expected to identify a specific process (1992:98). This mutation of the original configuration will also emit electromagnetic energy back out. In the photodissociation process, the molecule severs the chemical bonds and produces a neutral product. The small amount of the absorbed wavelength energy that is utilized in the creating of a neutral product is the quantum yield. Photodissociation occurs at all possible wavelengths that have a lesser energy value than the least amount of energy required at the state of threshold between unity and dissociation.

Although the photoionization process involves the ejection of an electron, a photodissociative product may still be formed. This free electron may then combine with other atoms or structures accordingly. The energies required for photoionization are usually higher than those of lower energy that requires photodissociation to occur. At shorter wavelengths where the energy is greater and ionization is more likely, this is an important factor to consider when applications of upper atmospheric molecules are concerned (1992:99). Changes to the tissue starting from that atomic alteration will invariably produce most undesirable results. Measurements of the ozone and oxygen components and their relative volumetric ratios to incoming electromagnetic energies are certainly of value.

The identification of ozone concentrations in our atmosphere can be accomplished by measuring the backscatter radiation from the ozone molecule. Since ozone absorbs the wavelengths of between 200 and 350 nanometers, a minimum reading of approximately 250 nanometers is expected to be seen from space (fig. 10.7). Ground levels reduce the wavelength over a specific wavelength, and for measurement purposes a mercury light source having a resonance at approximately 253.7 nanometers provides an adequate reference. While the ozone problem is made of two areas, one being the contaminant that is breathable and the other being the absorption of ultraviolet radiation, the first is considered a bad form of ozone and the second a very desirable form of ozone. The atmospheric measurement mechanism relies predominantly on the capacity of the ozone molecule to absorb within the band regions of Hartely at 200 to 310 nanometers for profiling. The Huggin's band region of 310–350 nanometers (1992:211) is used for total ozone composite.

Other ultraviolet absorption methodologies are useful in the determination of atmospheric gases such as the O_2 molecule, the oxygen molecule, and atomic oxygen (O). All absorb some ultraviolet wavelengths, and its spectral reference data can also be extracted from the sun or another light source such as an ultraviolet star. The most desirable method for ultraviolet measurement in reference to O_2 is occultation, using the eclipse of some solar body is used such as the sun or another star. The highest absorption measurement is usually a point where the path of its optical reference is closest to the Earth from the point of observation. N_2 uses an airglow method whereby during collisions or through a chemical reaction the generation of another product occurs that is in an excited state. When the products return to its preexcitation energy level, photonic emission occurs that can be sensed by high-altitude platforms arrays. Atomic oxygen (O) can be determined by the airglow method as well, using the emission band of 135.6 nanometers. The method of fluorescence is utilized by remote sensors to determine nitric oxide (NO), atomic hydrogen (H), and atomic helium (He). This method relies on the immediate emission of photonic energy. In these cases, the photon is the same wavelength as the absorbed energy; this is also known as a form of resonance scattering.

There are many different types of Earth measurement platforms, and the devices they employ to measure, image, and visibly observe the atmosphere are extremely varied. The information presented here is a general methodology of data received and internally manipulated through optical instrumentation, producing a measurement value that is compared to a known blackbody for electronic interpretation and then telemetered to Earth for compilation.

Not all weather-related synoptic imagery is processed from orbit using electromagnetic radiance as the primary source of energy. A development during World War II has provided the meteorological observer with a most accurate imagery of water vapor, volume, and surface direction (pressure gradients of highs, low, warm and cold fronts, and synoptic imagery) and uses a different approach to observable radiance measurement. This particular instrument transmits electromagnetic energy into the atmosphere from a known surface location and then measures the reflected and

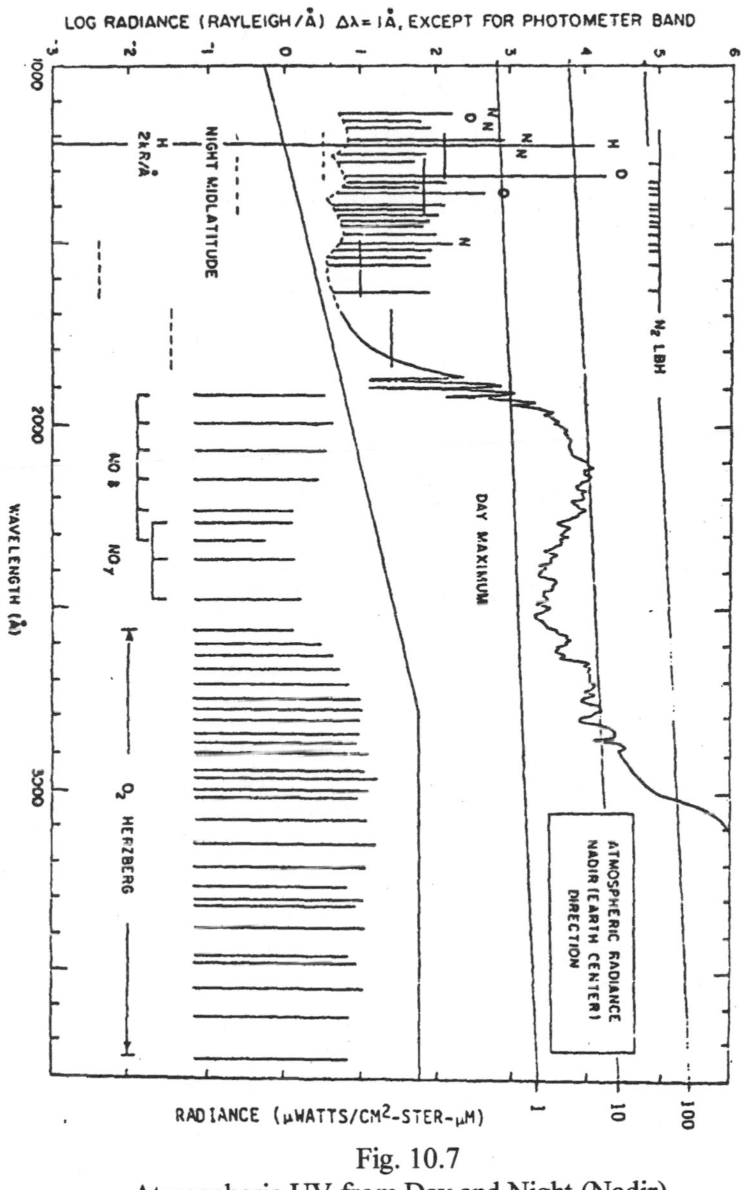

Fig. 10.7
Atmospheric UV from Day and Night (Nadir)
(Atmospheric Ultraviolet Remote Sensing:Permission:
United States Air Force

refracted electromagnetic spectrum emitted back from atmospheric molecules by their desecrate quantum processes. This invention is called the Ranging And Detection Analog Receiver.

Detection from Wavelength Energies

The equipment discussed here detects molecules and atmospheric particles as well as ranges them, or, put differently, determines how far the detected particles are from a known point of reference. This equipment is called RADAR, termed by S. M. Taylor and F. R. Furth of the U.S. Navy, made official for use by such agencies in November of 1940 (Doviak and Zrnic, 1993:1). This method is based upon the capacity of molecules, either suspended or as part of a solid substance, to reflect and refract electromagnetic energy detectable by a receiver tuned to the frequency of the reflected energy. This was suggested by Nikola Tesla in 1900 when he described the echo heard from one's voice. He theorized correctly that the voice heard was electromagnetic energy being bounced off the material, also that the electrical energy representing the voice traveled some distance to reach the material, which in turn reflected it back. This was the first simplistic method of describing reflected energy and still applies today. In 1856, Heinrich Hertz studied this property and demonstrated that at certain electromagnetic wavelengths objects would indeed bounce electromagnetic energy back to the originating source (Sauvageot, 1992:1).

Reception of electromagnetic energy reflected back from an object, determined to be from a specific object, would meet the criteria for detection of that object. That demonstration was recognized in Germany on May 18 in 1904, by the detection of riverboats at the Hohenzollern Bridge at Cologne, Germany, when they passed within an electromagnetic beam of energy at the wavelength of forty to fifty centimeters (Doviak and Zrnic, 1993:2). Although this demonstration was valid for the detection of large objects, the technical considerations outweighed the reality of its usefulness and for some years this concept was lost.

The next approach to using electromagnetics was during the year 1935 in England. The reasoning was not to detect an object but to incapacitate its abilities was first presented by the director of research at the Air Ministry. Mr. H. E. Wimpriss approached Mr. Robert Watson Watt, superintendent of the Radio Research Station of the National Physical Laboratory, asking about the possibility of producing enough electromagnetic energy in such a way as to incapacitate aircraft. His answer was based upon the wattage (power) required and the fact that if the aircraft was metal, a shielding effect would protect the crew (1991:7). Because the answer was most negative, showing that such an apparatus was not possible, Mr. Watt approached the request from another point of view. This viewpoint was essentially to detect aircraft and not necessarily destroy them. A proposal was drafted with this theory in mind and was submitted to the Air Ministry in February 1935, titled "Detection and Location of Aircraft by Radio Methods," which embodied the use of electromagnetic energy directed to engage and in turn be reflected by the aircraft upon its illumination. The reflected energy would then be captured by a receiver. I consider, that because the memo was circulated to the Scientific Survey of Air Defense, consisting of the distinguished professors P. M. S. Blackett, A.V. Hill, H. E. Wimpriss, and A. P. Rowe, its relevance was not lost within the possibilities that it must have developed among these men. On February 26, 1935, a Heyford Bomber was successfully detected when it flew through the radio energy of a BBC transmission. A team from the radio research station was immediately dispatched to Orfordness, and five weeks later the team of Wilkins, Bainbribge-Bell, Bowen, Willis, and Aircy successfully detected a flying boat at a range of seventeen miles. Later they demonstrated that a potential existed for illumination of aircraft and subsequent plotting of that object could be accomplished at a range of 100 miles. RADAR was born and the headquarters for that research commenced at Bawdey Manor (1991:8).

The RADAR effort was ostensibly utilized for the defense of England and through that requirement came the necessity of detecting and ranging hostile aircraft. The detection of aircraft was accomplished. However, part of the operation of detection is the fact that since aircraft were operating at tropospheric altitudes, imagery of those objects (aircraft) suspended within that envelope may be obscured from the constituents of that very atmosphere. For the atmosphere in general to obscure objects that bring destruction to the homeland was of great significance, and studies of the atmospheric field were undertaken to understand and eliminate such echoing possibilities (Doviak and Zrnic, 1993:5). Although studies were undertaken to modify detection capabilities, not until the 1940s did the components required to produce wavelengths that would be short enough and at the output power high enough to satisfy precipitation echo reflections exist. J. T. Randall and H. A. Booth were involved in the research of this problem at Birmingham University in England and eventually resolved this situation by combining the resonant cavity feature of the klystron with the power capabilities of the magnetron cathode, creating the final configuration that is now the basis of a contemporary magnetron. With this development, higher frequencies (shorter wavelengths) were available that would enable the imagery separation between large objects and atmospheric particle echoing. In July of the year 1940, radar operations by the General Electric Research Corporation Laboratory in Wembley, England, were involved using a ten-centimeter wavelength. One of the research scientists, Dr. J. W. Ryde, was conducting tests on the identification of echoic properties that stemmed from atmospheric clouds and rain. Although evidence directly establishing this work is not conclusive, most likely due to the utmost secrecy that surrounded the project and its findings, the fact that Dr. Ryde was indeed involved with this research does identify these years as active in a pursuit concerning meteorological applications of electromagnetic detection capabilities (Doviak and Zrnic, 1993:5).

Early radar imagery in the 1940s was initially an illuminated spot on the graphical surface of a cathode ray tube, similar to a spot in the center of one's television set. The representation of an object that had been illuminated would grow in size on the cathode ray's surface as it approached closer to the transmission location. This ballooning effect of display size was the only reference that the observer had to determine range. With the development of shorter wavelength transmitters, the ability to project the distance (range) of the illuminated object from the observer became possible. The transmission energy would be turned on, then off in a pulsed fashion. The pulsed transmitter along with improvements in receiver technology provided the necessary bridge establishing target detection and ranging. In the case of meteorological radar applications, the pulsed method allows for the atmosphere field to be detected and ranged; not only the precipitative properties but also meteorological targets from radar energies can now be utilized to identify the atmosphere itself, that is, the clear (nonprecipitative) atmosphere or particle medium (field). Both precipitative and nonprecipitative information regarding the field in whatever condition is exposed in parameters of velocity, thermodynamics, hydrologics, isobarics, profile, and consistency. All these variables will eventually lead to discoveries that enhance the accuracy of hydrometeoric and particle behavior.

Because the pulsed transmission radiating outward from the source is reflected back and since the target is fixed in location, the wavelength of reflection will be the same at the receiver. However, if the target moves because of a velocity shift, the scattered or reflected wavelength back to the receiver changes. Should this occur and the receiver is not looking for any other wavelength except the value transmitted, a ranging plot is not possible. This effect is called a Doppler Shift. Operators were aware of this physical phenomenon during the war when they received echoes from the object arriving back at the receiver at the same time as radiation bouncing back or from the scattering effects of other objects (Doviak and Zrnic, 1993:5). Therefore, to properly understand why radar can detect and plot the ranging of atmospheric particles from their velocities using the Doppler effect, some information on the basic concepts of Doppler should be explored.

Doppler Shift

The term *Doppler* originated from the descriptive nature of how the propagation of electromagnetic wave changes in relation to an object in motion, either moving away from a certain location or moving toward the location. The first explanation of this phenomenon was described by an Austrian physicist named Christian Doppler (1803–53 (1999:143) in 1842. To understand the Doppler explanation, if you are standing near a road and a car horn is blown on a vehicle that is traveling toward your location, you will experience the sound waves from the car's horn increasing in frequency. If the car is moving away from you and the horn is blown, the frequency of that horn will decrease as you are listening to it. This is the physical manifestation of either a compression or elongation of electromagnetic energy. In figure 10.8, this relationship between an object's relative position to a known location can be seen to influence the reflective energy accordingly.

The preceding paragraph will apply to hydrometeoric situations whereby a typical raindrop is not static nor is a volume of such objects and therefore would be subject to motion, hence velocities. Since the transmission carrier is of a pulsed type, the scattered return would also be pulsed, and hydrometeoric velocities would affect the pulse so that its frequency phase angle would be altered accordingly. Figure 10.9 will identify this particular sinusoidal property as it phase shifts according to either an approaching or receding target. The difference in frequencies and their phase

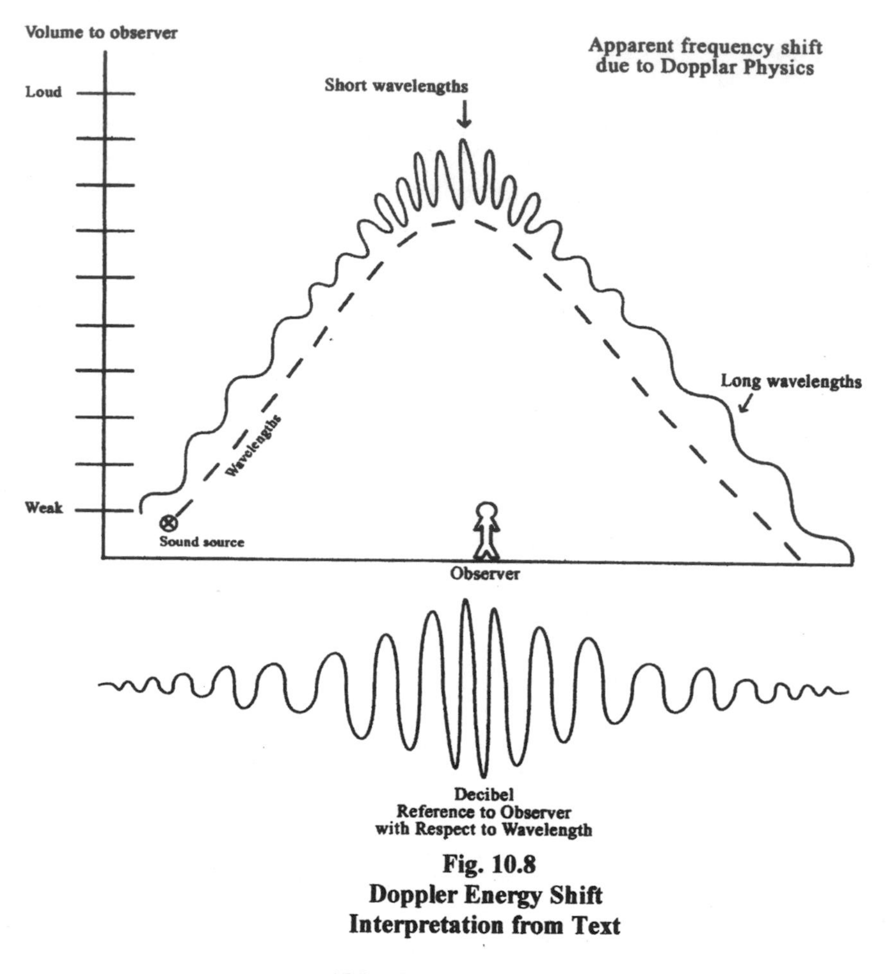

Fig. 10.8
Doppler Energy Shift
Interpretation from Text

(Mendes - Hussey Graphics)

angles accordingly identifies the Doppler shift characteristic, and this difference is the echo that is measured. The microwave transmission in pulsed format will have a much higher frequency that the few kilohertz identifying the shift; also, the wide band capability of the receiver in general would have a greater capacity to see phase angle shifting without any trouble. Because the transmitted pulse was a frequency with a specific phase angle relationship, that particular value in hertz must be remem-

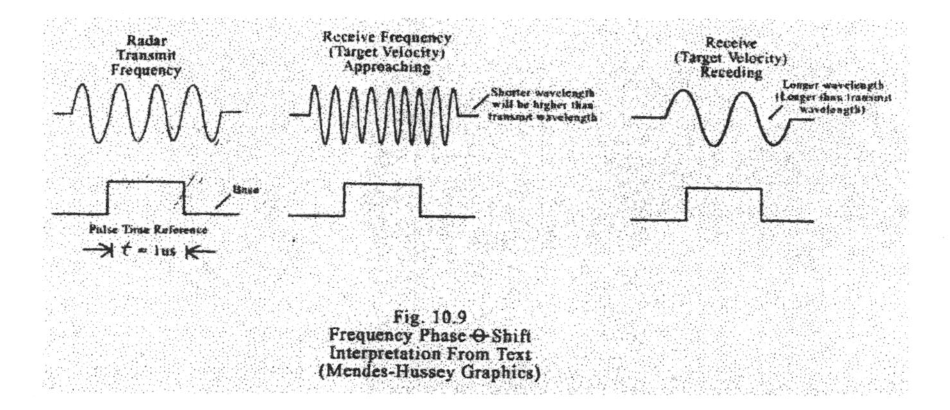

Fig. 10.9
Frequency Phase Θ Shift
Interpretation From Text
(Mendes-Hussey Graphics)

bered when it comes time to do a comparison analysis to the received frequency for the purpose of an identification that satisfies Doppler Shift. This remembered frequency is provided by an independent oscillation obtained from a local oscillator termed a *STALO,* or *Stable Local Oscillator,* designed within the component circuitry (Sauvageot, 1992:9).

Target ranging and velocity data are not compromised by the methodologies of the electronic mixtures requiring multiple frequencies generated internally to the circuitry. When two frequencies are mixed, the resultant frequency is still sinusoidal in waveshape and therefore any shift characteristics would be sinusoidal as well and any shift direction ambiguous at best. A phase demodulator circuit eliminates this problem through the mixing of both waveforms with the reference waveform being advanced or having a slight phase angle offset. At the demodulation output, the signal will be zero if the frequencies are the same. If the value is positive with respect to zero, then the target has a velocity approaching the observer, and if the demodulation output is less than zero, the target has a velocity away from the observer (radar location). The generalized block diagram of a pulsed phase coherent radar can be seen in figure 10.10.

Quantitative data representing the atmospheric medium is generally defined at about one gigahertz for the transmitting pulse frequency. At this high wavelength the median will deliver accurate pictures of its transparency, hydrometeoric content, velocity, and temperature. The scattering content of volume and its measurement physics can also be observed. Atmospheric quantums as far as reflectivity are concerned are considered to comprise the oxygen and water vapor interactions that would be interac-

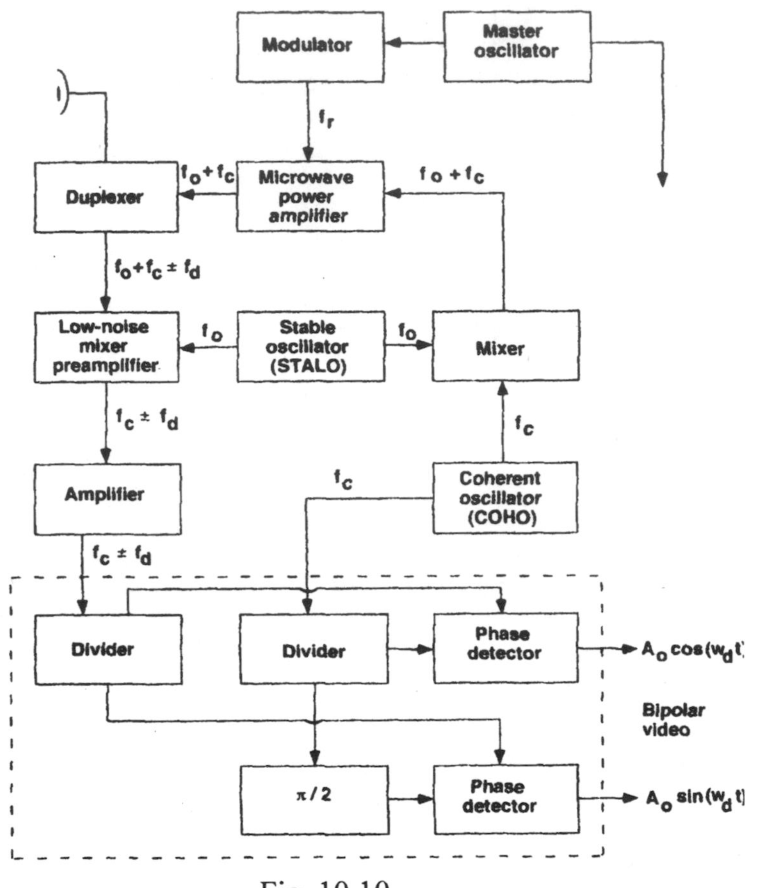

Fig. 10.10
Coherent Pulse Phase Radar Diagram
(Radar Meteorology:1992-10)

tive with the wavelength energy. Meteorological radar also enhances the quantitative analysis of scattering particle size, absorbitivity, and both precipitative and nonprecipitative forms of cloud configurations. Clouds themselves will constitute a myriad of particle sizes with a boundary limit between precipitative elements occurring around two hundred um in size (Savageot, 1992:70). At this hydrometeoric size, the fall velocities would be .75 meter per second to the minus 1 power (.73m/sec[-1]) and correspond to a millibaric pressure of 700 or an altitude of 3,000 meters. Upon growth, the particle will no longer stay in the shape of a sphere but elongate to an approximation of the ellipsoid. Current physical data show that the spherical size limitation for large raindrops is six millimeters; however, this data is always subject to modification as more information regarding precipitation and its growth and breakup becomes available.

The volumetric distribution concerning clouds indicates that the concentration using maritime air will have only a few score large droplets within a centimeter cubed, versus landmass volumes that have considerably more smaller droplets and a much higher count per centimeter cubed (Sauvageot, 1992:77). When considerations regarding radar echo and backscatter properties are concerned, droplet size and quantity per medium volume are important. Understandably, echo and backscatter effects are much greater when hydrometeor sizes are larger. For calculations regarding this property a general distribution analysis using an exponential quality of R was developed. This value offers a realistic approach to identifying a rule-of-thumb distribution pattern through the function of hydrometeoric diameters for the value of R (fig. 10.11), *N* in the figure represents melting particle distribution.

Other parameters that radar transmissions and their subsequent images that they present regarding a meteorological target are provided by the microstructure itself. and that cannot be derived easily with linear systems. With a polarized emission the physical information representing shape and particular orientation within the medium can be resolved (Sauvageot, 1992:123). In a linear polarized energy beam, components that are horizontal and vertical to the emission path of propagation (h) and (v) are vectorially defined and represented by a flat plane.

Circular wave radar emissions, as opposed to straight line linear, is an encompassing field that is generated in two halves about the propagative path.

Because a raindrop has a tendency to form into an ellipsoid during freefall, due to the flattening effect as a product of its water volume by acceleration and gravity forces, the equivalent volume increases and, because of the new geometric ellipsoid configuration, with regard to the cross sectional backscattering property, is almost logarithmic. This characteristic is imaged showing the drop displacement as a greater equivalent volume diameter.

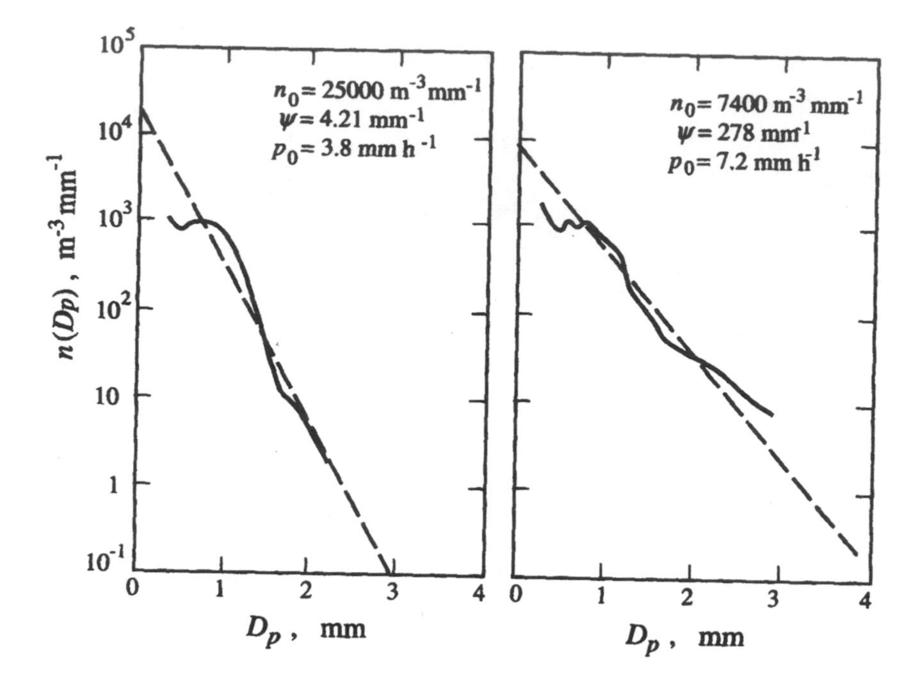

Fig. 10.11
Rain Drop Diameters with respect to Distribution
(Atmospheric Chemistry and Physics:1998-832)

Polarimetric Doppler

This is the name of a particular emission used to image rain. Because the falling characteristics of raindrops are different from those of cloud mediums, beam transmissions must also be different in order to properly interpret the variations of echo reflections. To fully understand this radar enhancement, some advancement differences between a linear Doppler transmission and a polarized Doppler transmission should be explored.

To understand the concepts of circular radar, it must be noted that the essential difference between the two is the circular field that is generated (fig. 10.12). The left circular and the right circular fields are pronounced, with the right circular field being in clockwise rotation as the propagation

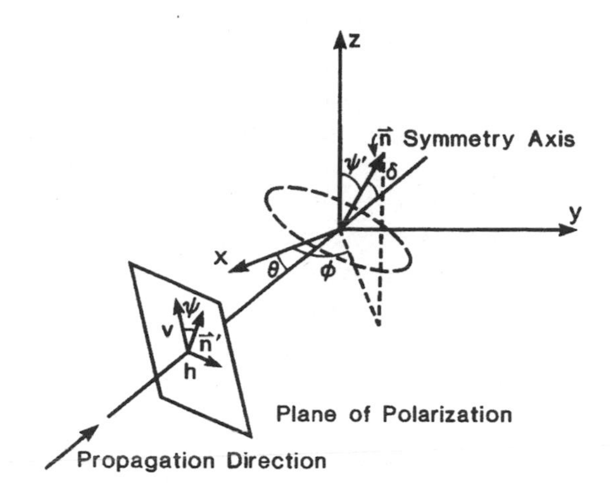

Fig. 10.12
Scattering Geometry for n,h and v
(Dopplar Radar and Weather Observations:1993-250)

of the energy wave moves away from the observor. The other generated field is opposite in rotation.

In linear polarization, the radar transmission is radiated outward and will comprise separate vectors in both the horizontal and vertical component . The term for this transmission is *LDR (hv)* or *LDR (vh)* or *Linear Depolarization Ratio* (Doviak and Zrnic, 1993:267).

Because of the reflective characteristic of an ellipsoid upon a Polarimetric waveshape, the returning wave that is received is also elliptically polarized due to the phase differences between the right and left polarization rotations. In an h or v polarization (linear) waveshape, the canting relationship will be zero (the received polarity is equal to the transmitted polarity and there will be no depolarization) if the ellipsoid has its axis parallel or perpendicular to the propagative plane of geometry or its axis is not equal to the polarized plane, then the backscattered wave will posses the same orientation as the illuminated ellipsoid If the drop is spheroid, and is radiated by a circular waveform, one of either the left or right channels will be missing and the diagonal terms equal zero. In the case of an ellipsoid under radiation from Polarimetric waveforms, the phase of the two left and right components will be determined by the ellipsoid orientative geometry (125).

Radar is used to differentiate many other physical phenomena such as hail and lightning. When depicting lightning, the reflectivity of the signal is due to the free electrons that are dissociated during the ionization of the atmosphere by the electrical discharge. Although a signal of backscattering can be obtained, the time element of its duration is quite small because of the recombination of the electrons with other matter particles within the atmosphere after the discharge. In a basic sense it is the amount of free electrons within a specific volume that determines backscatter intensity and duration from cloud to ground situations. In a relationship of wavelength to time of observance, most observances will be made with wavelengths greater than 10 centimeters with a detection window of 100 milliseconds after the initial discharge. Also, since echoing may accompany the observed discharge and be imaged, specific wavelengths using a circular waveform are appropriate. In addition to the lightning, echo imagery upon precipitation may also be affected due to the polarization consequences that such high temporary electrical fields have on the orientation of hydrometeoric properties. Thunderstorm densities will affect radar wavelengths as the bending of that energy is distorted and intermingles with normal backscattering.

We have covered just a minute part of the applications that radar can be utilized for. In the imagery of backscattering, many situations can be discerned. Wind velocity, forms of both precipitation nonprecipitation, as well as cloud identification and atmospheric temperature, can all be resolved through the fact that atomic matter reflects energy in many ways that can be received and interpreted accordingly. Fundamentals of radar concepts are only discussed here, and through this general information regarding the extraction of atmospheric information from reflectivity and absorption of matter further interests can develop. Technology in the interpretation of backscattered energy will define an invisible situation that may be many miles away, and through the imagery developed from that reflectivity eventually comes the actual atmospheric chart that was started by defining the medium in Miletus by the Atomists.

The Charting of an Atmosphere

From a standard concept that a chart identifies attributes that provide knowledge, then there are many forms of charts that exist. The blueprint for a home is a form of chart in the details that it presents to the builder. It identifies internal knowledge necessary to complete the areas that must be installed. Electrical, hydrological, structural, and financial aspects all must be given in such detail as to create a workable unity. An engineering diagram for bridges and buildings and electrical and mechanical designs convey information required to fabricate in material form a product that conforms to anticipated attainment specifications. Although the information regarding weather in itself may be formatted to identify a specific occurrence, the differences here are that the occurrences are being diagrammed for their detailed internal and external behavior. The chart of engineering entails the knowledge to create and not to analyze the existent object or behavior. That analysis is done after the creation to verify specification compliance. The differences between meteorological and electrical, architectural, or a creative specification chart are therefore between a behavioral consequence and a dedicated product with an expected performance.

In addition, since the atmosphere is a constant variable, outcomes from planetary influences will continually modify the observance and demonstrate the need for continual surveillance. In essence, a home or manufactured product will be the same an hour from now; a thunderstorm may either dissipate or become a tornado.

The need to create some form of quick identification for a meteorological occurrence started in earnest during the years Mr. Henry worked with the map that was placed in the Smithsonian. From that beginning, the historical nature and significance of thermal and hydrologic manifestations concerned with our atmosphere was placed in the forefront of the scientific community.

From the earliest records, charts were first associated with land and the constraints evident for travel to and from those bodies. As empires expanded, the need for access to those lands was kept a secret from the general populace, due to the perceived importance of transit in the acquisition of territorial and geological wealth (Brown, 1949:121). The meteorological aspects were descriptive regarding the wind direction and cloud formations that would either assist or hinder the expeditions of trade accordingly and, only reflected primary information necessary for direction and distances. Ancient writings that identified any mention of the winds themselves, were described from the direction that they came from, and their association to anything beyond good or evil was not considered important. The Greek philosopher Aristotle stated that the wind does not have vertical motion, and only four winds were accounted for. Homer's writings also stated this observation, and the writings of the Bible also state that the Earth contains only four winds (Revelation 7:1). Later the Greeks established in their writings twelve winds, and in the sixteenth century the writer Georgius Reisch also mentions them, except without their Latin nomenclature. The Latin rose of the twelve winds existed during the Roman times and was still in use through the Middle Ages (Brown, 1949:125) (fig. 10.13). Today the modern rose in a combination of both the magnetic bearing and the winds.

The weather map specifically published for the general public's reference with realistic meteorological standards was first associated with the newspaper and initially began as a combination of two important aspects with regard to a presentation of public issues. The issues relevant were those of the day, which would be normal and consistent with journalistic applications.

First of the two parts to the presentation was information that would be issued in due course as would be in a general approach to the populace, with the usual verbiage to generate emotional interest. The second reason for meteorological inclusions was cast in the light of importance to public affairs, with the focus of presenting information targeting society and the commerce that relied upon weather prognostics. The early nineteenth century was a realization that the costs associated with meteorological publication within a daily time frame required issuing maps from the Weather Bureau as a method of increasing the paper's readership circulation, and from that revenue increase came the method for generating the offset costs that were required for the total publication, which had originally been taken from financial allocations that had been set aside for the gathering of that applicable data (1999:153).

The original printed daily weather map itself was a product of England. The first published map appeared in the *London Daily News* on April 14, 1865, as a compilation of the meteorologist William Marriott's (1848–1916) efforts in association with data covering a two-month period of twenty-nine English cities and the cooperation of the Royal Observatories superintendent for magnetism and meteorology, Mr. James Glaisher

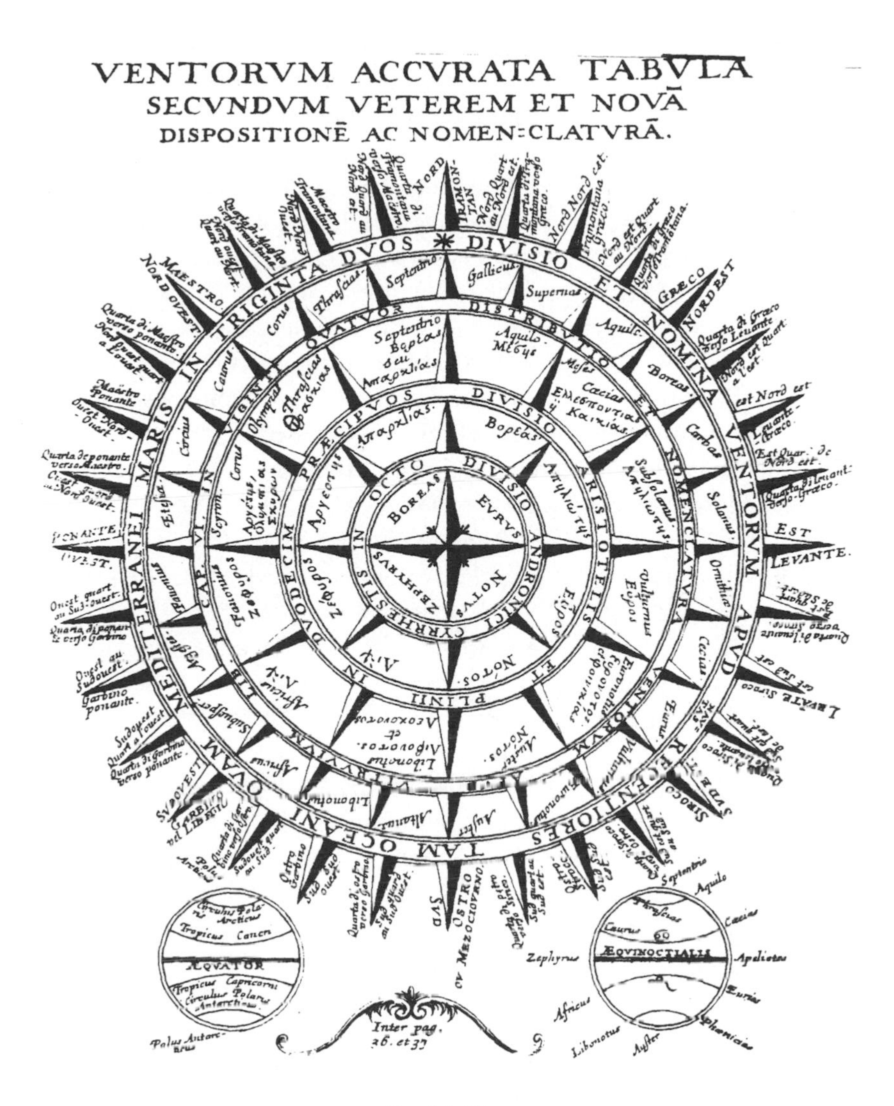

Fig. 10.13
The Four Point Rose Evolution from Homer to the composite Rose in the 17th Century (A History of Maps:1949-134)

(1999:154). The map is shown in figure 10.14 and depicts the weather as it was reported on April 13, 1865. It was Mr. Marriott who gave the *Daily News* the first credit for a synchronous observation, which was published in June of 1849.

It is important to observe that this map is a depiction of weather that had existed at the time of the reporting and does not represent a forecast. However, this was a truly significant step in the overall picture of bringing meteorological information to the readership. Later the *Times* added a second edition with reports that were taken from the telegraph stations at 6:00 P.M. With this new information, the Meteorological Office stayed open later to facilitate the newspaper's requirement of producing a second map. In 1879, two more papers followed suit and assisted with the cost of telegraph operations to facilitate their information-gathering quest as well. The two additional papers were the *Standard* and *Daily News.*

In 1897, this time on May 9, came the first continual regular publication of weather information from an American newspaper called the *New York Daily Graphic* (1999:158). Because of the size of the Unites States, information that was transcripted to the newspapers was centered in Washington, yet the papers themselves were not. A cipher or coded message containing the plot information was telegraphed to the Weather Service observer in New York, whose job it was to prepare the weather chart as a facsimile or one that represented the chart that had been prepared in Washington that morning. The pressure and temperature plot points were sent via telegraphy in this coded format, which looked very similar to the bingo references we use today. The coded format was compiled from a grid network that was placed over the completed weather map, and the points that each parameter represented were issued with the grid system coordinates by which they were encompassed. This method essentially increased individual interpretation and was the chief reason behind any lack of accuracy in the placement of meteorological data, and associated errors were possible anywhere within the extent of the individual grid dimensions.

Symbology

In the definition of information's means, some standard of presentation must be in place prior to its dissemination that will provide a coherent language of fundamental understanding. In the symbology of telegraphy, it

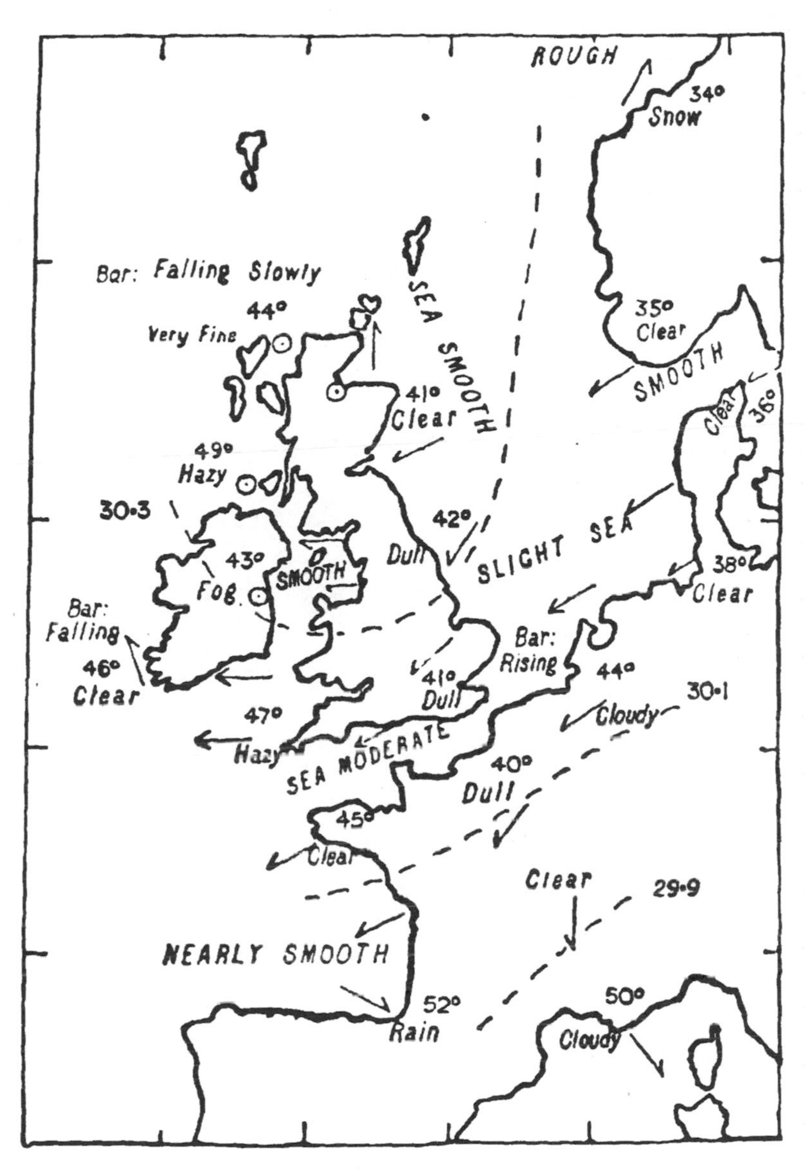

Fig. 10.14
The London Times Weather Map-8am, 4/14/1875 as reported in the London Daily News
(Air Apparent:1999-155)

was the physical arrangement that the Indians used to communicate, not the flags or cloth. The placement of arms about the body of the signalman determined the communicative format. In this way, a language could be developed using the alphabet as a reference to literally spell the communication. In the early years of meteorological presentations, several informative and innovative representations explaining atmospheric occurrences were developed. Since the weather maps were published for a general readership, the contents inscribed over each map tried to reflect as much meaningful data as could be inserted. Readerships that were within circulation localities interested in outdoor activities could extract temperatures that would assist them in any planning that required the possibilities of cold or heat. Other weather maps extracted pressure gradients with wind identifiers listed and showing their direction. One particular piece of information that stands out among the diverse issuances throughout the nation was that in almost all cases, the characteristic of the atmosphere was listed as consistent with five different phases of behavior. In addition, an example of this (fig. 10.15) identifies clear, fair, cloudy, rain, and snow, with each map done differently in such manner as to reflect the individual artist's preferences for writing and presentation style. Map differences beyond style were attributed to variations in litho and print medium because of technologies available at that period. Differences in ink, paper, and the metallurgical aspects of the type set all contributed to the overall problems across the nation with respect to the advancement of meteorological symbology. Only a portion of regional areas were for the most part consistent with the paper's circulation. East Coast papers and West Coast papers presented the United States generally as far as the Dakotas topographically drawn; however, for the Midwest, information presented was sketchy at best. As the years passed, from 1891 to 1909, stations issuing their own version of weather maps increased from 52 to 112 (1999:164), and this alone created the necessity for a more standardized map that could be amended at the local station once they received the Weather Bureau's model. Sometime in 1910, a policy from the National Weather Bureau under the Department of Agriculture was issued by Mr. Willis Moore, the bureau chief at that time, stating that all station maps internally drafted would be discontinued in favor of a commercial map that was to be made available from the Weather Bureau. This map represented the total United States, with standardized symbols identifying pressure in solid isobaric format and temperature in isothermal representations. The map listed all contributing stations with local capability to amend their own wind and atmospheric data.

The evolution of the weather map contained many diverse areas that required a language compatible with the advancement of meteorological studies. As universities and other school curricula evolved through research and a general interest in this physical science, the representatives of that academic learning upon graduation were becoming an integral part of the communicative world. This included the telegraph stations and eventually newspapers and, in our more recent world, wireless communication. A network of stations having meteorological departments was fast becoming part of the accepted communicative process in many parts of the United States, and interests regarding the natural phenomena of our atmosphere were not only additional but also authentic. Media needed to adapt to another form of satisfying the readership's interests in this matter, rather than the exclusionary written accounts containing only social, governmental, and business affairs within the circulation and country as a whole. People were beginning to take interest in what was around them atmospherically, while basically the only information available was in a written language that many without reading skills, let alone in English, found difficult. This particular physical science was an innovative way of presenting the atmosphere around them that all could experience regardless of academic or social standing. For all the diverse ways that many people may understand it, only the graphic structure was expedient. This provided the mechanism that was required in order to overcome printed ontological philosophies. Instead of a paragraph that was language- and philosophy-driven, the picture of a symbol now proved to be the best way of understanding of that data, as long as the symbol was communicated in such a manner to eliminate interpretations. A specific symbol dedicated to the fundamental understanding was needed to convey that a certain behavior was pending as the result of an atmosphere occurrence. The symbol eventually designated for that purposes also had to be founded in sound physical law.

A graphic structure needed to represent an understanding should first be recognized as legitimate in what its message is, so that a corollary can be established between the symbol and expected event. There must also be a clarity of understanding that depicts the specified event in that it (the symbol) must be understandable to a diverse populace in academic and lingual philosophies. With these rules applied to this effort, the consequence hopefully would offset the variety of possibilities that each individual interpretation might suggest.

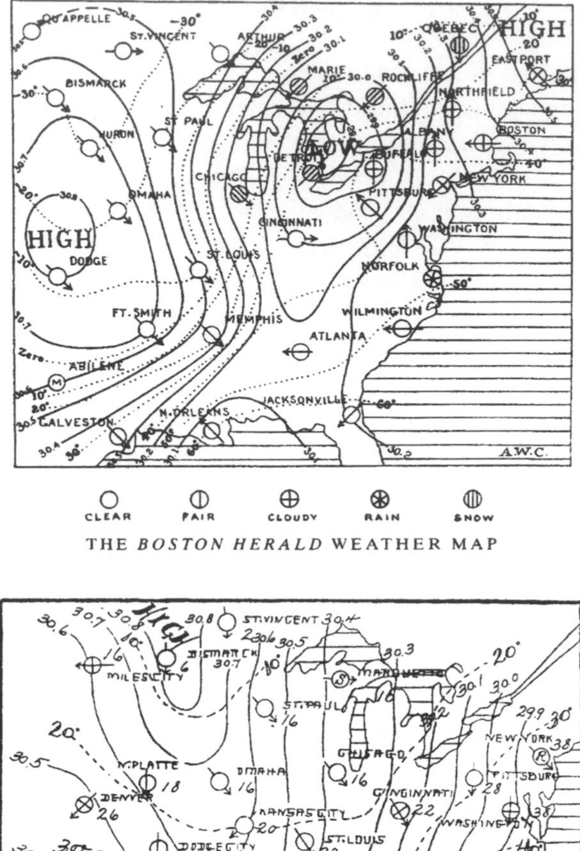

CLEAR FAIR CLOUDY RAIN SNOW

THE *BOSTON HERALD* WEATHER MAP

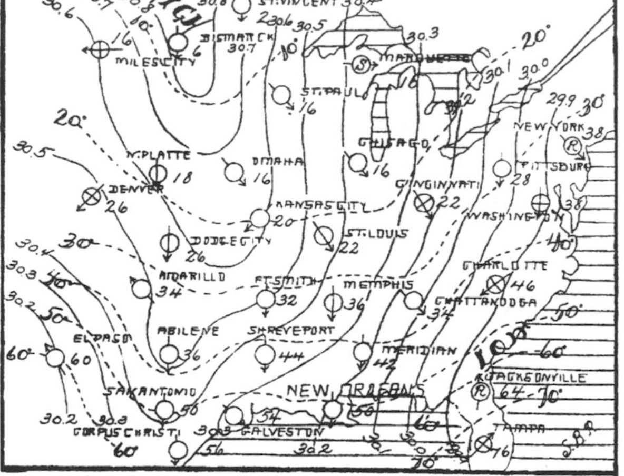

Explanatory Symbols.

Clear, Partly Cloudy. Cloudy; Rain; Snow.

THE *NEW ORLEANS TIMES-DEMOCRAT* WEATHER MAP

Fig. 10.15
1894 Weather Maps
(Air Apparent:1999-162)

With regard to its dissemination, whatever is to be established as a definition in symbolic form must still remain clear enough after it has been sent to the variety of stations that will use it and would have to survive a multimedia effect of copying and electromechanical reproduction elements once the original had been received from the Weather Bureau. Although the symbols themselves once chosen by meteorological principle can establish the mechanism that conveys a rudimentary understanding for their message, they can also indicate to the general populace that additional disciplines of learning or prompting would be required.

As a case for an example, the symbol currently used to identify a low-pressure center is a large *L,* yet the average individual without a meteorological background or a fundamental understanding of Boyle's Law in physics would no doubt misrepresent the meaning or not know that it identifies a recognition that reflects change. In this situation, the symbol is just one alpha character and simple at best, yet its meaning is far more comprehensive. This can be a substantial problem when simplicity is used to convey a more complex association with physical law.

A sunny day, however, requires very little beyond a basic recognition that the sun is visible without clouds in the sky and it's daytime. In addition, the observer should have a basic knowledge of what a cloud would appear to look like in the sky. These particular associations with atmospheric occurrences are the simplest and are learned at a very early age because they are part of everyday life. Early in the history of weather association, its identification with any passage of weather was generally confined to traditions and placed within popular and general observances. Examples with ants, cats, flies, and any number of variety of animals were used to identify meteorological occurrences of which the animals themselves seemed to have an awareness. The altitude of birds, the congregating of bees, the thickness of a bear's hide, the skin of onions in the garden, et cetera (Chabaud, 1994:134, 135), all had simple rhyming phrases that were generally acceptable in the identification of pending weather patterns. The early weather maps and charts issued to the populace needed a different way that would express pending weather without the necessary requirement that one be able to read. Certainly a weather map would be preferable to phrases within the paper's circulation that told you to observe your cat's movements or barnyard motions from the livestock. Therefore, any association with the sun and its obscurity by clouds in its simplest form would require a symbol that was easy to relate to and would be understandable to a majority of readership.

Representations

The sun in the simplest depiction to most anyone would be in the form of a circle, with gradations of sunshine represented as lines drawn through that circle. Notice here that the circle alone without consideration of its inside boundary is a clear day, one line bisecting the circle signifies partly cloudy, an *x* through the circle represents overcast, and a circle that is blacked out or shaded would mean periods of storm and rain. Meteorology itself, however, is not as elementary as just the presentation of a simple circle, because complex patterns and expectations from synoptic levels require additional symbols that need to portray a greater complexity and understanding. Barometric pressure, wind direction, temperature, and the patterns that clouds have by what they represent with regard to those parameters are certainly more complex, as we have now come to realize through the observations of their motion, direction, and formations.

Is a sunny day as simple as a weather front? Which begs the question, What's a front? More important, would an individual care about a weather system that would have progressions through many complex phases, instead of knowing just four representations of a circle and their meaning as would be applied to a clear, partly cloudy, overcast, or rainy day? The answer lies in the fact that the four phases of a day, from clear sunshine to rain, are relative to expectations of immediacy; at least, to the individual that might mean the current day and can be dealt with by a simple observation of the sun. However, more complex atmospheric functions present expectations from learned definitions that require higher levels of understanding and a deeper comprehension of meteorological science. In these cases, many perhaps cannot grasp the higher order of comprehension or they might not posses that particular scientific discipline that is necessary to convey the understanding.

If representations of meteorological manifestations are to be understood well enough to perhaps anticipate an occurrence anywhere from six to thirty-six hours from the present, how are the details of that advanced meaning accumulated to eventually appear on the weather map for the readership to ponder? An explanation of the meteorological representations would be required to inform the general readership of those representations used and to assist in the interpretation of a more complex relationship in the designation of that educated thought.

Typically, the Weather Bureau receives many transmissions from other observing stations across the country, and when one message is received on the teletype machine at the Bureau others connected to that same circuit around the world also receive that information. There are many teletype machines connected to different networks, each identifying perhaps local, regional, national, and international originators as well as law enforcement, highway, and the media of newspapers, radio, and television. All contribute to this center local information that is relative to meteorological aspects of weather systems across the United States.

Figures 10.16 and 10.17 will be used as examples to identify weather pattern phenomena typically used in meteorological presentations. For the most part, each reporting station will produce a specific array of meteorological data that will conform and presented in a standard format, as seen in figure 10.16. This data is typical for observations taken and reported by that station. Meteorological reference to all pertinent pressure, temperature, cloud density, dew point, wind direction, and some tendency over the past six hours of parametric variances are also displayed. The reporting stations are then placed with their respective map locations and connected with additional synoptic presentations that would be of concern with regard to frontal systems moving across the country (fig. 10.17).

In the analysis of figure 10.16, one of the most illuminating pieces of information presented is the barometric pressure. Since all measurements are related to sea level, connections between and from each reporting station presents a most interesting pattern. These patterns can be seen in figure 10.17 and are called isobars (refer to Chapter 6) and demonstrate the difference in millibaric pressure through the distance of their spacing between the lines. The spacing increments identify differential columns of molecular densities about their concentric points as either a low or high. Because columns of pressure are dependent upon temperature to commence convective forces associated with them, that vertical and horizontal motion is attributed to the variances in terrestrial thermal equilibriums, which determine the air flow in convergent or divergent dynamics. The gradient wind itself, that is, the actual atmospheric molecular density moving from a high- to a low-pressure source or specifically a change over distance through an acceleration process, would be considered parallel to curved isobars and generally located above the terrestrial friction layer at a higher altitude than the geostropic surface wind (Lutgens and Tarbuck, 2001:175). An additional factor in the consideration of wind velocity and direction is that the geostropic wind is a straight direction vector (not concentric) and has velocities proportional to pressure gradient differences,

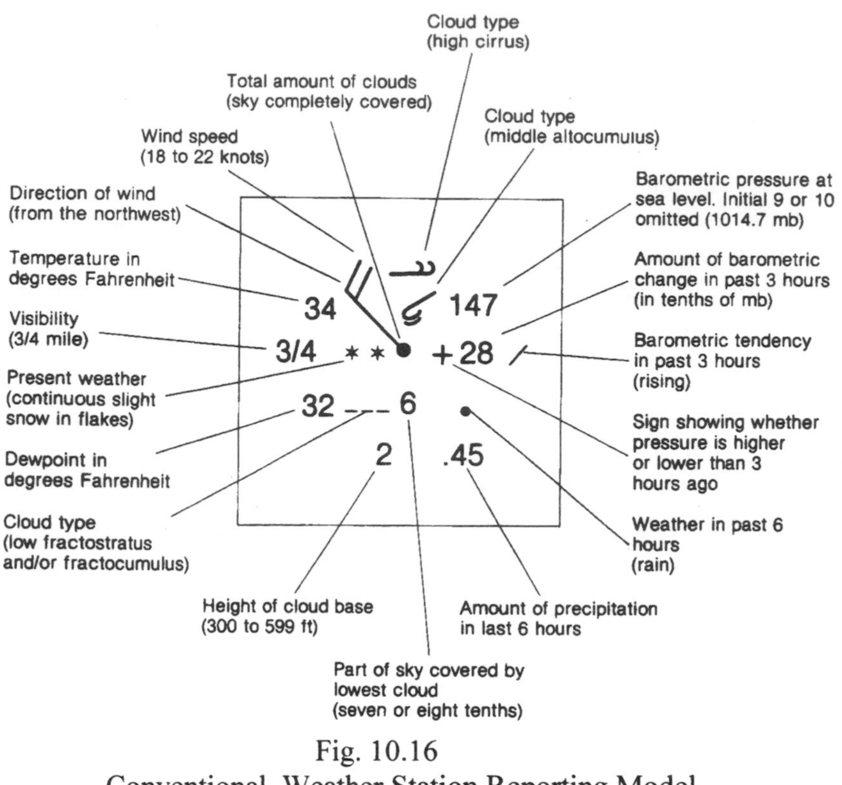

Fig. 10.16
Conventional Weather Station Reporting Model
(Meteorology:1991-413)

whereas a gradient wind is developed through frictional and centrifugal forces delineated by concentric isobaric locations and its velocity is constant as determined through accelerative forces defined as the horizontal pressure gradient (1991:227). These all assist in the determination of the pressure center and surface winds. As a result of the Coriolis force, surface winds identified on a surface map possess velocities in an inward direction, crossing concentric isobar lines toward the center of low pressure, and an outward direction, crossing concentric isobars about the center of high pressure (see the surface wind directions and their corresponding velocities in fig. 10.17 for an illustration. Also see figure 10.16 for a clarification of station parameters that identify this meteorological representation.) Further, pressure differences cause molecular densities to begin flowing,

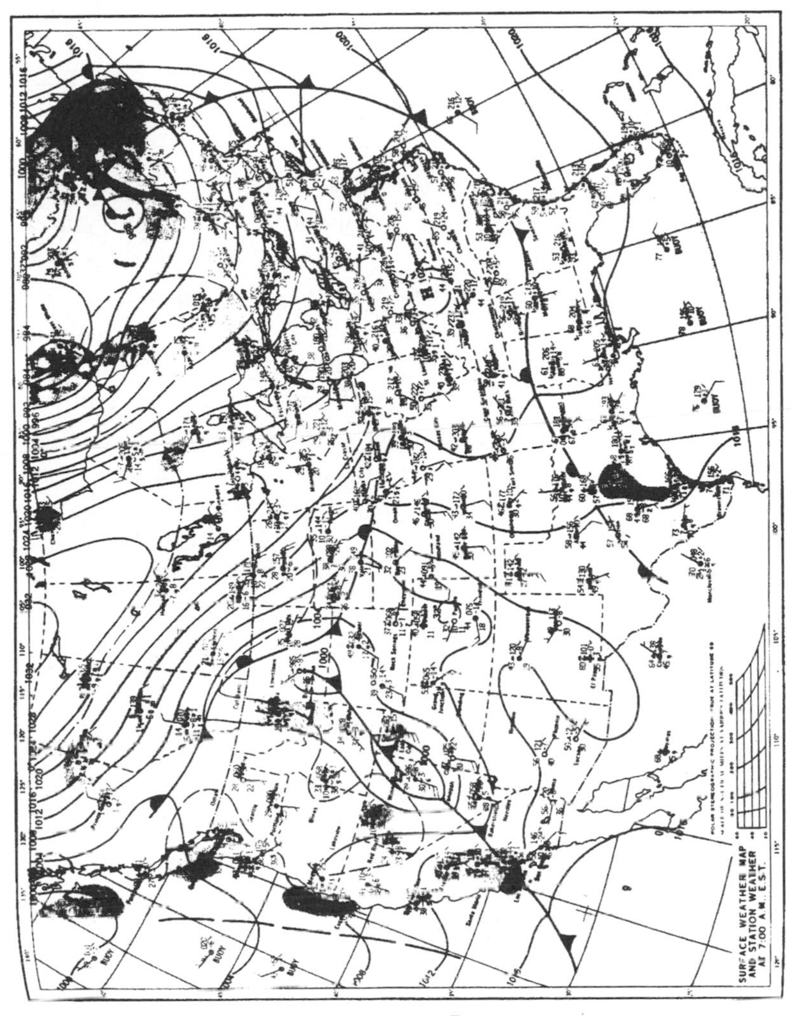

Fig. 10.17
Conventional Synoptic Weather Surface Map
(Meteorology:1991-415)

and these may be in a coasting situation, not under acceleration, and in a relatively straight line along an isobar (geostropic) until they come under an influence from a pressure center and begin to flow in a curved and concentric path (gradient wind).

The Front

To represent a difference in two volumes of atmospheric mediums (densities) covering a large and broad distance and that may or may not be in motion, the term *front* is used and associated with two types of symbols. The first is a line that is denoted with triangles along it. That particular weather map symbol represents a cold front. The line that has half-circles along it on the weather map denotes the location of a warm front. Although the differences between the two densities may vary in temperature and moisture, the basic definition would be between two air masses in general. The difference is usually about 9 to 120 miles in width, represented as a broad line (Lutgens and Tarbuck, 2001:259), with the aforementioned meteorological symbology along the front's edge. Each air mass (density) moves with a relationship to the other; that is, each mass may move either faster or slower than the other and through this velocity difference may produce an overtaking configuration. The term *front* was created by Norwegian meteorologists because of the similarity this represents to battle lines. At the meeting or overtaking point between the two masses, a mixing will occur; however, because one air mass may represent cooler air and the other air mass may represent warmer air, a displacement will generally occur. The cooler air mass if overtaking a warmer mass will penetrate the lower warmer mass's boundary; this is termed *occlusion.* In cases where the temperature of the air behind a cold frontal system is colder than the air in front of the system, that situation is termed *cold occlusion* (Vasques, 2001:52). In the reverse situation, where the temperature of the air is warmer behind the cold front than the air temperature of the air mass ahead of the warm front, the term is *warm occlusion.* If a warmer air mass overtakes a cooler and slower air mass, then the warmer air will ride above the cooler, denser air, and this is called *overrunning.* In either case, it is always the warmer, less dense air that will rise above a cooler displacement mass. These descriptions of warm, cold, and occluded frontal systems are illustrated in figures 10.18, 10.19, and 10.20. I use the word *system* because each frontal boundary has multiple parts to it that assist in recognizing them for what they are. Also, because each frontal system is different, they determine an expectation precisely because of their configurations in advancement, which is represented by the formation and procession of specific clouds.

Not all frontal systems are in motion all the time, and in many cases a

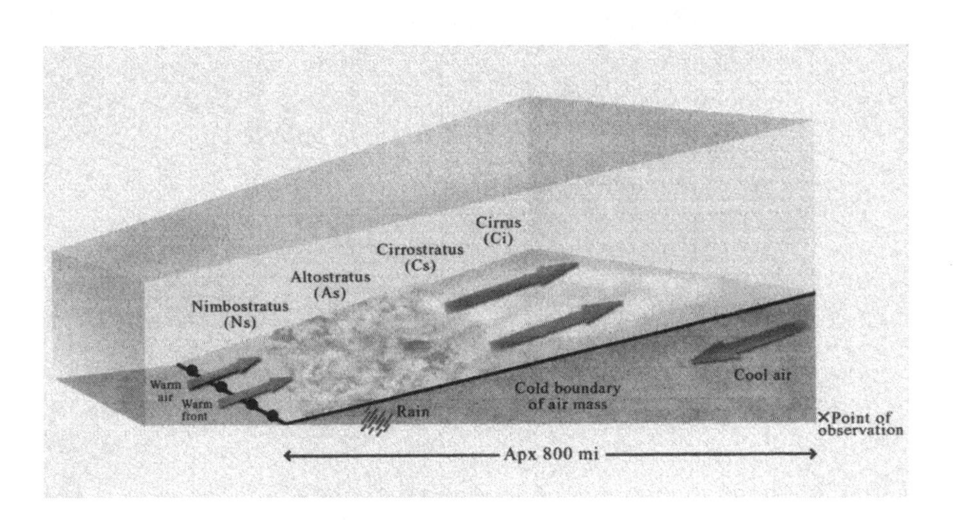

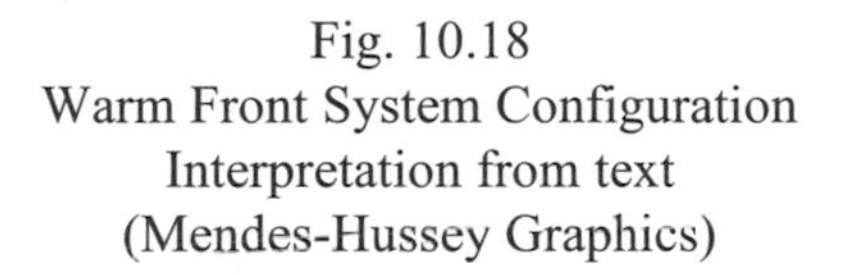

Fig. 10.18
Warm Front System Configuration
Interpretation from text
(Mendes-Hussey Graphics)

front will remain in its location for a prolonged period of time; the term that describes this situation is *a stationary front.* In cases such as these, the precipitation associated with them may last for several days and will lead to excessive rainfall or snow. The advancement of any frontal system can usually be determined by the cloud formations preceding the moisture. In addition, the air temperature and barometric pressures associated with the passage of these fronts will also show a marked difference. In Chapter 4, a study of Boyle's Gas Law was undertaken to establish the foundation for reasoning behind this transition in the physical atmosphere. Because pressure, density, and temperature are all related when associations with frontal variables are considered, the individual mechanisms that they represent in atmospheric outcomes with their passage as an example will buttress the principles of Chapter 4.

Since pressure is the product of the multiplication of density, temperature, and a constant, pressure can be stated to be the product of just two variables. Although volumes will invariably change, temperatures may not affect that temporal and spacial consideration. When air temperature is raised, the air density may not increase with it. If the air temperature falls, the air density may not decrease with it. In the case of winter, the temperature may drop as surface pressure rises. Specifically, when discussions

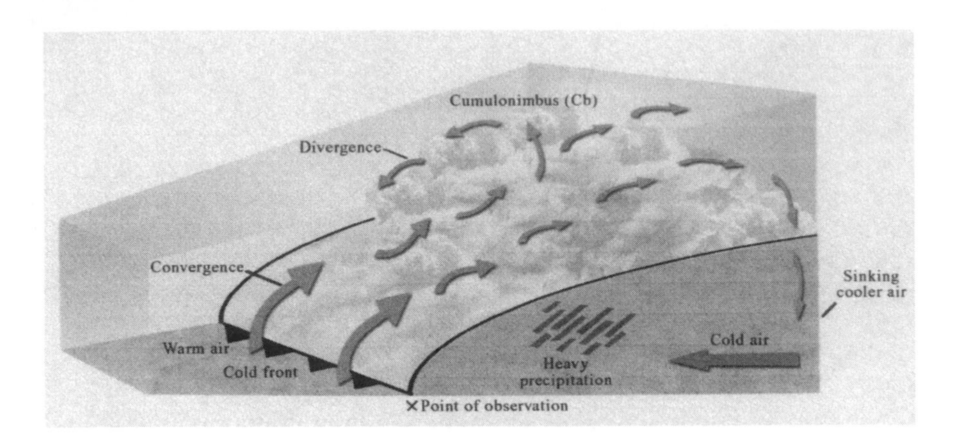

Fig. 10.19
Cold Front Configuration
Interpretation of text
(Mendes-Hussey Graphics)

evolve around low areas of pressure, if a gradient wind flowing into the center of a low pressure is entering more rapidly than the air can rise upward, both air density and pressure will increase. If the center of pressure is high and more air is descending than diverging at the surface, the density and pressure decrease. All should be considered when advection properties are involved and when expectations are made concerning the passage of frontal systems.

Some descriptions of fronts use the terms *polar, seabreeze,* and *jet,*; all have significant meaning and should be discussed. The term *polar* when associated with frontal systems usually refers to a very high contrast in temperature. In this particular situation, there is a specific delineation about the Earth at latitude sixty degrees. This is the general boundary of where the polar cold air reaches the warmer westerlies blowing from the latitude of about thirty degrees. A polar frontal system would normally be found in the regions of Canada and South Russia and would extend downward to about the southern tip of Greenland. An Arctic high is very cold and stable air that is a small cell that breaks free of the polar air mass and will travel into the United States in a southeasterly direction. In some circumstances, this very cold air (high-pressure cell) will reach Florida, where the impact on citrus growers will be severe.

In the case of air moving from and into warm and cold areas, *advection* is a term that identifies this dynamic. When a pressure differ-

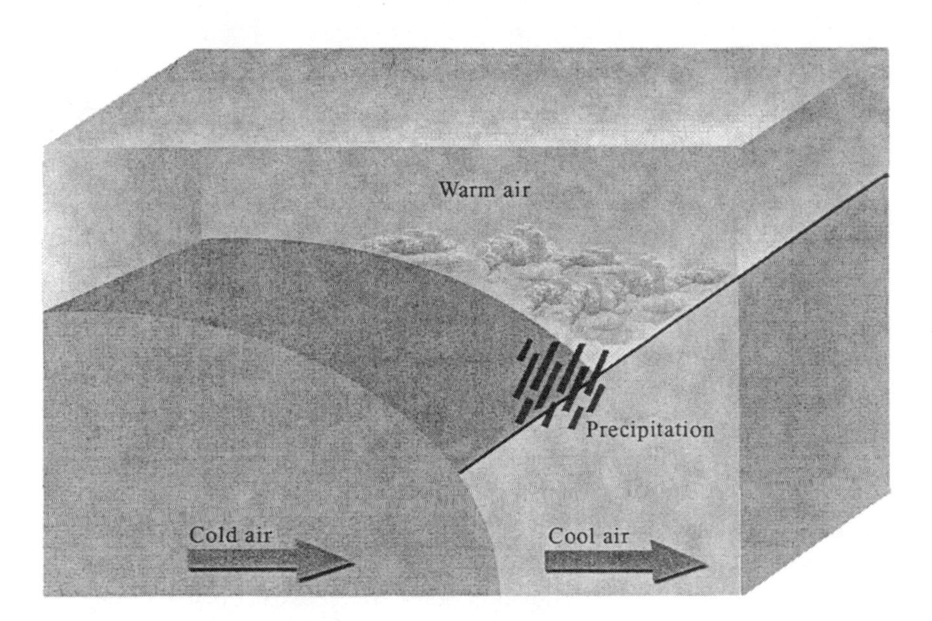

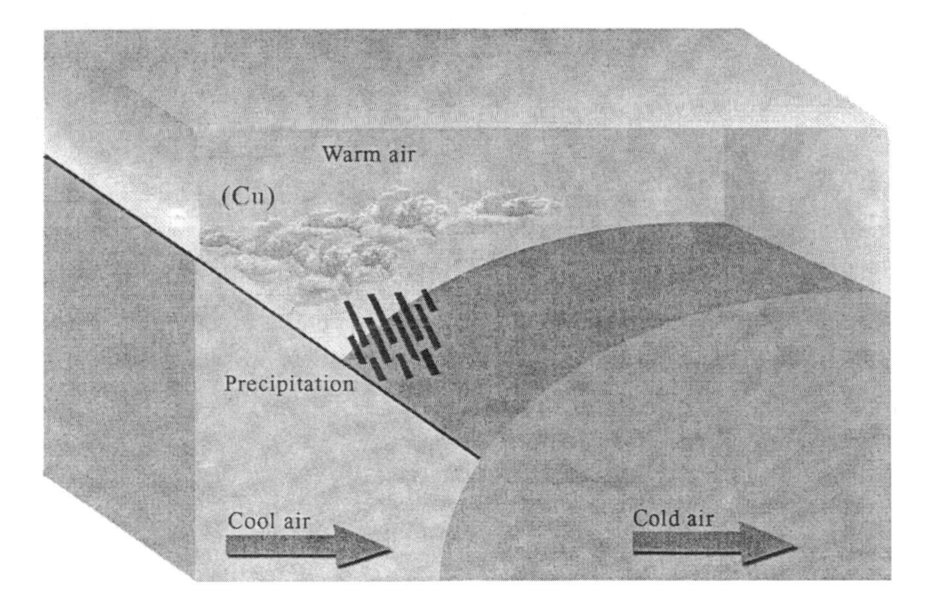

Fig. 10.20
Warm and Occuluded Frontal Systems
(Mendes-Hussey Graphics)

ence causes colder air to move into where warmer air was, that is cold advection. When warmer air moves into a region that was cold air, that is warm advection. The term *jet,* as it implies, means "fast wind" and represents a very narrow band of extremely high wind at an approximate altitude of thirty thousand feet. The meteorological term is *jet stream.* The regions of this ribbonlike band of wind can be seen in figure 10.21 and are not readily definable about the Earth. When the land meets the mass of water, the jet stream is particularly high, developing wind speeds of almost 273 miles per hour (Danielson, Levin, and Abrams, 1998:1001–264). Because the jet stream is aloft at altitude, its dynamic motion assists in divergent outflowing, which creates the development of cyclonic activities lower nearer the planetary surface. As the jet stream moves from left to right across the topographical surface, regions above and to the rear or north will develop convergent flows; the upper right of the jet stream or the region north and to the front will develop divergent manifestations of convective flows. The opposite will develop below the jet stream. The region to the rear and below will be divergent, and the forward half of the jet stream and below it will develop convergent properties.

The jet stream will also develop seasonal shifts, which will aid in its strengthening during winter months when the temperatures are the greatest in difference and lessening in the summer months, when the temperature differences between the northern and southern United States are less. As a general rule, when the jet stream is seen to be moving across the northern parts of the United States on a weather map, warmer air can be expected to occupy the Atlantic region. When the jet stream is moving up through the southern United States in a northerly direction, then an expectancy of colder air in the Atlantic regions will follow. The summer jet stream is located about the latitude of the Great Lakes; however, this shifts as winter and the Earth tilt, forcing the jet stream to move lower across the southern United States. This obviously causes the winter air temperatures in the northern latitudes to decrease as a consequence and assists the Arctic air mass to propagate into Midwest regions as well.

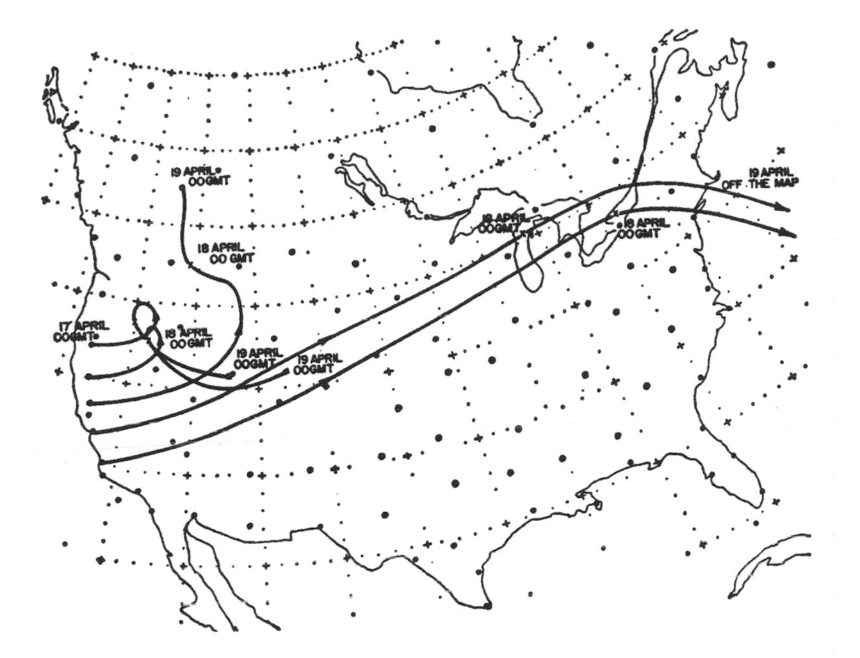

Fig. 10.21
Typical Synoptic Jet Stream Configuration
(Jet Stream:1967-82)

11
The Words in the Sky

The Forecast

Can a prediction be made that will be somewhat accurate in the forecast of weather? Can anyone observe the sky on a particular day and say that rain or other precipitation will occur three days later or tomorrow or that day? Yes.

The sky is a variable that identifies fluid motion in the way it appears to us when it is observed from the ground. In these observations, there is an apparent sequence that seems to follow certain cloud formations that promote an expectancy of weather. That weather may be in the form of an overcast sky, windy weather, or clear and warm or rainy and cold. What is important here is that the clouds observed in the sky at certain times will indicate a period of weather that can be expected to follow with a certain degree of accuracy.

Not all individuals will be carrying or have in their possession meteorological instrumentation that will readily display barometric pressure or the temperature, yet these instruments for all their values will only portray the current situation relative to the placement of that instrument and will only display that information data at that time for interpretation. To consider the readings as data that foretell impending weather requires continual monitoring of that data and a plot of some fashion that will indicate a trend. Looking at the sky is relatively easy, and the clouds or lack of such combined with the wind factor if one is available will lead to a fairly accurate picture of weather that follows. The messages that clouds portray can be of significance to an individual with a basic understanding of atmospheric trending.

Cirrus clouds (Chapter 6) will indicate the upper-elevation wind directions, and because they form in the altitude of jet streams, their ice particle relationships to following sequences of variable lower density cloud formations are very important, because the lower cloud formations that follow these "mares' tails" configurations will certainly arrive accompanied by winds that will blow from a much different direction altogether. Here is a relationship between high-altitude cirrus and the lower wind that identifies a different direction source when observed from the surface. Because of the wind direction and its relationship to high- and low-pressure sources, its direction at the surface can be much different from that of the geostropic gradient that follows isobars at higher altitudes and is in the region of those clouds (cirrus). At the surface, the gradient wind is experienced by us and its relative direction to cirrus observations can be most significant. Because of these physical observations, a basic rule can be established. With your back to the surface gradient wind, the observance of higher cirrus clouds moving from your left side, will mean that the current weather situation will deteriorate (Watts, 1999:70). If a correlation is made between this statement and perpendicular surface wind observations and the fact that cirrus clouds identify a warm front approaching, the wind at the most leading edge of that warm front will be almost ninety degrees to it. So if your back is to the surface wind, the higher cirrus edge of the approaching warm front will be moving at ninety degrees to the surface wind, or from your left side (fig. 10.18). Associated with the warm front's passing will be an increase in the air temperature. At the time of the observation witnessing the high cirrus formations, expect rain or precipitative occurrences within the range of twenty-four to thirty-six hours. The high cirrus (fig. 6.1) formations will gradually be replaced with a lower covering sky (stratus) of altostratus, and altocumulus (fig. 6.5) clouds will be observed. When the altostratus or thicker cloud formations replace the cirrostratus, the precipitation will be approximately twelve hours distant. As the lowest cloud formations of the warm front eventually arrive, those formations that carry the actual surface precipitation will resemble nimbus and nimbostratus (fig. 6.8) structures and the time element for precipitation will have decayed to approximately five to eight hours.

An additional cross winds rule of thumb states; if an individual stands with his back to the lower surface wind and high cirrus clouds are moving from his right side, the weather will improve (Watts, 1999:70). That is to say, expect a cold front to move into your area, which generally indicates a more stable atmosphere. Because of this temperature drop, convection properties will subside and cloud formations will dissipate, leaving a period of blue sky, until the cold front barrier is fairly close. That situation will be evident when tall cumulonimbus formations appear. These formations are due to the warmer air that is convergent to the front barrier of cold air near the surface rising over that cold barrier and with that convection causing condensation at higher altitudes, which gives the cumulus formations their thunderstorm and billowy appearances. One of the reasons that a period of blue sky or clear sky will appear is because after the passage of a warm front the ground is typically wet because of the rainfall. This situation will cause a lack of condensation and the liberation of heat until a period when the ground is sufficiently warmer to allow thermals to properly convect (fig. 10.19). With these indicators, rain associated with cold fronts is usually heavy and will converge upon the observer very close to the frontal barrier of high cumulonimbus formations. These types of fronts may bring lightning and large thunder squalls that include hail because of the nature that such high and convective forces within them influence. The altitude of the cloud structures allows for subsequent electrical stratums to develop within the cloud where those configurations generally produce a difference of electrical potential between the cloud's base and the ground, causing discharges between the two.

In the previous paragraphs the discussion was about the fronts in general and the precipitative qualities that each would produce in a rather expected time frame. With this as a background in forecasting, the decision of the kind of precipitation is left. Expectations of rain, snow, ice pellets, rain/snow mix, or freezing rain can now be exposed to determine if the forecast for precipitation would foretell which kind will fall.

In review of the various books that cover forecasting in general, the basis that many use in the determination of precipitative kinds is the altitude of upper-level systems. Typically, the expectation of rain can be made if warm air extends to at least 2,000 feet. By *warm air* is meant here "above freezing." With the upper-level atmosphere continually above freezing, any precipitation that occurs regardless of either a warm or cold frontal system will invariably produce rain at the surface.

If warm air aloft overruns a cold front (fig. 10.18), several situations as a consequence of this situation will occur that need careful focus, especially during the time of the year in winter latitudes. First, warm air aloft at or above 1,200 feet will develop the presence of liquid (Vasques, 2001:105). Previous studies will confirm that occurrence. With this in mind, then the cold air mass depth is the determining factor as to what the water drops will develop into prior to ground impact. If the cold air altitude above the surface is less than 800 feet, the water drops will not have time during their fall to freeze into a more solid lattice. The expectation here would be precipitation in the form of freezing rain at the surface level. If the cold layer is greater than 800 feet, then the raindrops will form into a more solid lattice because of the time spent within the colder air. This situation would call for the expectation of sleet.

If the warm layer of air aloft at altitude is less than 600 feet thick and fairly negligible, the occurrence of any rain penetrating the cold barrier staying in a raindrop configuration is small enough that snow can be expected. As rain is reconfigured to a more solid lattice structure, such as snow or ice, and it passes through a warm layer of air and begins to melt, that activity will extract heat from the atmosphere around it and the warm layer will start to diminish in volume and/or intensity. With this occurrence, the surface will experience a change in the form of precipitation that will undergo a transition from rain to sleet and then, eventually, to snow.

Summary

Atoms that encompass our world can be modified and bonded together in a variety of molecular relationships providing for our existence, to include not only the breathable format but also the configuration of water. This statement contains a great amount of comprehension, so it is that with any heading that states "Conclusion" or the derivative of rendering an opinion indicating the understanding of philosophies and therefore some foundation of supporting evidence should be presented.

Particular references in this case can be found at the back of this book, where sources representing the body of my presentation throughout the text may be reviewed. Facts taken from the reference are presented within a specific ontology, enhancing the information extracted from the source and supplied in a different method of written format. I choose this to present knowledge in a way that may be more comprehensible, which in turn I hope constructs an awareness within the reader.

We now understand that the foundation of the word *atmosphere* itself was an explanation of the medium where birds fly, originally stated over 2,000 years ago, and became a reality when that understanding was presented around 300 B.C. Defined as a substance consisting of indivisible matter called atoms, these incredibly small particles were the reason behind the resistance individuals experienced during times of wind and was imperative to the mechanism of breathing. Because the exposition of atomic matter was an important revelation such a long time ago, the fact that it could not be seen would still continue to gave way to the speculation regarding its existence. This, however, was not to last, and as the centuries passed and the deification of knowledge moved into the more empirical men of a scientific persuasion replaced pseudoscience. In the 1600s, matter that constituted the invisible form around us was given the name gas, and through compression came the irrefutable evidence of its existence. The 1700s finally completed the invisible puzzle with the presentation by Lavoisier that gas was not just of a singular atomic composition but of multiple components, consisting of both nitrogen and oxygen. In 1808 a pivotal moment occurred when Mr. Dalton defined the small particle that had been called the atom as an "element." He also postulated a theory that, in basic physics, stated elements could adhere to others of their own kind and there was more than just the elements of the air.

The early 1900s structured the atom more specifically with presentations by Rutherford and Bohr that gathered knowledge into the actual configurations of the atom. Eventually a stratum of elements were created that defined their hierarchy based upon the number of orbiting electrons and corresponding protonic numerations each element possessed. Because the electrons in elemental configurations are susceptible to outside stimulus by way of other energy particles or quanta, they will react in some cases, altering the parent structure they belong to by either moving closer to the nucleus or leaving the structure altogether. In addition, with their response is the possibility that energy will be released by their reactiveness, which may be equal to or different from the quanta the electron received. This is a fundamental method of creating or altering the molecule, which is comprised of atoms itself.

Within the atmosphere are molecules that have the configuration of one hydrogen atom and two oxygen atoms. This is a molecule of water, and properties that exhibit its dynamic are part of our everyday life. Other molecules that have a great meaning because of their reactiveness to quanta are ozone; however, when discussions are relevant to the atmosphere and its principal behaviors, the water molecule is of higher significance.

With the exposition of this molecule, discussions centered around the morphosities that the water molecular can undergo when exposed to thermal stresses. The molecule can be excited through an interaction with thermal influences, causing reactions that define a dynamic in gaseous, liquid, and formations that are rigid. Morphosities in the phase changes from the process of sublimation to deposition offer visible proof that this particular molecule is a priority when considered in terms of our existence. Further analysis of the water molecule identified the various formations that become visible when its molecular structure is exposed to other physical properties that develop within the encompassing gas. We see these formations as visible structures that appear to float above and around us, and we have termed them *clouds,* with each type identifying a certain expectancy in its arrangement and subsequent progressions to those of other categories.

Instrumentation was created in the sixteenth century that made it possible for us to actually observe the physical properties that the atmosphere has in its relation to earthly forces such as gravity. With these observations science added a new dimension to the atmosphere, which later became defined as density; consequently, this discovery led to the conclusion that the atmosphere had weight, thus exerted pressure on all matter. This physical property was verified through a new instrument that had been developed and given the name barometer. Further inspection of the density gave way to the fact that as gas the atmosphere would sustain suspensions of a variety of atomic substances, including the droplet of water. The formations of that suspension were defined in the late 1800s into seven classifications by Luke Howard and later updated by Abercromby and Hilderbrand into the categories we have today.

No discussions that illuminate the atmosphere can be undertaken without the effect its densities have from solar bombardment. Photodissociation is an important aspect of any atomic density, because of the irradiated properties that are particular to any solar photonic discharge, which constitutes a definite significance for terrestrial matter. Dissociative properties from solar ultraviolet radiation will affect the O_3 molecule to produce the oxygen molecule O_2, which is the molecule that we breathe. The absorption of that particular radiation is almost 100 percent by the ozone molecule, and it is because of that absorption property that the harmful ultraviolet rays are withdrawn from the atmosphere around us.

Just as the radiation emitted from the sun would be most harmful if not for certain molecular components, the charged electron particles that flow through the molecular field around us will also protect living species living on the surface. The Earth possesses an electron shield that is defined by their motion and therefore defines a magnetic field that exhibits a dynamic characteristic from the south pole to the north pole, with each particle moving in such a way as to compose a form of highway that plasma particles from the sun's high-energy emission will follow. This magnetic field is part of our atmosphere and plays a vital role in providing for the passage and conduction of those high energies to the planetary north pole.

The influence is its principal protective element. It shields lifeforms from the harsher reality of other particles that possess a defined electrical orientation and energy value. Either way, the matter that is defined as atom and its associated electron play an important part in the creation of an influence around us, and that is the essential foundation that we know as atmosphere.

The early 1800s expanded the territory of the United States with the addition of newly purchased land, and it was because of this that science had to render adequate communication to those distant parts. The Lewis and Clark expedition was just the tip of the iceberg, because it produced the final connection between new territories and the weather that moved from them to the east. This was an important step in the promotion of any communicative methodologies that eventually materialized as a result of hostile weather and its descriptive requirements. In concert with electromagnetic advancements using copper wire conductances, several companies established forever the fact that information concerning weather and the destructive nature of such meteorological phenomena was imperative and important to the general populace. The data gathered from distant territories could be plotted in such a way that its presentation to the general populace became a daily occurrence; this established a format. With the eventual placement of observation posts situated in diverse areas, the weather picture became more synoptic in scope and a clearer and more defined pattern of motion dynamics could be discerned. This gave way to the conclusion by Mr. Espy that areas of differences in atmospheric density produced the movement of wind.

Climatological data was deemed so important that the surgeon general's office received reports on a daily basis concerning pressure and wind. This led to the foundation of the Smithsonian Institute in 1855, which provided a more centralized point of disclosure, and although this institution eventually discontinued its practice in this area, the importance of weather and atmospheric intelligence became formal in the creation of the U.S. Signal Office in or about 1874. Instrumentation and communicative inventions propelled atmospheric studies to a point in the 1930s that its control was eventually placed within the Department of Agriculture. Continuing scientific achievements in optical platforms led the way for the eventual placement of the TIROS orbital platforms that currently provide such vast amounts of visible real-time surface dynamics presented in today's media and have led the way to introducing some form of forecasting ability.

With optical advancements, the spectral associations of radiative properties that water possesses can now be imaged in a false color. That is, specific emissive and absorptive radiation properties can be observed that are not within the band of visible light. This aptitude produces a greater unmasking regarding the internal workings of water-borne formations, which could not previously be deduced using the naked eye, especially from such distances as a terrestrial orbit. Greater effort within the infrared and its associative spectral wavelength studies have greatly increased our knowledge of water particles through the observance of their behavior as a result of their dimensional characteristics. Refractive, scattering, and reflective observance abilities are enhanced by geometry and size, which manifest additional measurements as they react within an electromagnetic field. This then paved the way for optical science developments that initiated the impetus into behavior research, leading to explorations of how solar radiation is influenced by interactions with cloud and general atmospheric particulates.

Further enhancements in electromagnetic theory influenced the possibility of not only detecting spectral emission but also ranging the source of that emission. This study became known as RADAR and is utilized today to determine the density and distance of meteorological phenomena. The synopsis entailed a brief history as well as the advancements from its inception in the early 1940s to the phased array that incorporates emission characteristic known by today's nomenclature of *Doppler.* Quantifiable meteorological data can be ranged more accurately with higher frequencies, and since certain molecules and atoms will be energized into a reactive state using shorter wavelengths of energy, this makes targeting specific chemical particulate and particle configurations more easily defined.

In Chapter 10, historical and logical increments beginning from accepted definitions of the term *chart,* to current representations, including established symbology and descriptions, have been presented to further establish observable behavior communicated in a format that enhances knowledge and therefore interest in the general population Because of this fact, newspapers, other print media, and television, now have available to them the result of various established sciences dealing with meteorology, which provide the world populace with visual symbology and an optical clarity of identifiable atmospheric behaviors in real time.

In 1910 the mapping of remote data compiled through observer reports was commercialized to depict U.S. boundaries with symbols that replaced a great deal of philosophical presentation. Alpha characteristics made the dissemination of weather trending and observable formations easier and more readily acceptable within the general populace, producing the validation that rudimentary understandings were possible and could be comprehended by a majority of any readership. Eventually, with a media requiring a picture rather than the thousand words and the fact that television is primarily a presentation of graphical representations through audio and visual senses, a greater explanation of deeper terminology could be dispensed because of the associative nature of visual identification with an audio explanation. With this as a foundation of educating the general public on meteorological ontologies, the actual formations of hydrological manifestations can be presented with more synoptic extensions. Understanding of the air mass itself gives a greater range of forecastability because of the visible territory under observation from an orbital viewpoint. Most important, observations from a vantage point that is incredible from a third person's participation in data retrieval that requires such technological achievements allows the general populace to visualize their global location with respect to observable formations in motion and with this association can begin to determine the extent of their expectation.

Conclusion

I believe that the information presented within these pages has satisfied the requirement first set forth in the quest to chart our atmosphere. The definitions of the two words *atmos* and *sphere* were explored to determine the ontological roots of these terms. Greek anthroprogenic foundations can now be associated with the people who first involved themselves with the definition while living on the shores of the Aegean Sea. We can now state that evidence shows it was these people who uncovered and eventually created the science of invisible properties that, prior to instrumentation, could only be accepted through the most basic of experiences. Behavior was no longer associated with deity motivation but rather a physical existence that in some cases became exceedingly hostile. It was then from this understanding that matter existed in such a minute amount that came the effort to quantify that theory. This undertaking began the exposition that eventually became our periodic table and its list of atomic elements.

I have presented to readers the fact that the gaseous envelope surrounding our terrestrial surface is a mixture of atoms and through their individual quanta force establishes a coexistence with other atoms that may be different. An exploration was initiated to satisfy the definition for the molecule, and with the explanation of those structures the door was opened to their behavior. Our atmosphere also consists of particles that are smaller than the atom, and these tiny bits of matter may possess a difference in electrical potential between them that originates from distances far away from this planet. The energy that these particles contain is considered to be the radiation that impinges on the atmosphere above us. Some of this radiation is very lethal to many lifeforms, and the ability of specific atmospheric atoms to accept and thereby remove harmful radiative properties from our environment is a significant understanding in the benefits that the atmosphere possesses.

The science of meteorology is rather specific to physical explanations regarding mass motion and for constituent matter that represents an atmospheric structure; however, the ability to communicate behavior dynamics from thousands of miles away with such technological clarity had to be in place within the atmospheric science to accomplish this. Instrumentation therefore must play a significant role in any understanding that requires explorations of atomic and subatomic components, because we cannot directly observe their individual behaviors on such a microscopic level. In addition, the combination of technology applied with concern for human existence will always allow for greater achievements and accomplishments.

Wherever we go in our lives upon this planet or, for that matter, any off-world that contains an atmosphere, we will always be conscious of the fact that our very existence relies upon the ingestion of a gas. The constituents of that gas are then obviously very important to our survival. I truly believe that it is the importance of that reasoning that directs and gives energy to its interest.

I am in awe of what the atmosphere can do for us as a consequence of its behavior. We can enjoy a sunny day or we can experience extreme disruption to our everyday lives. The atmosphere is not something now that is misunderstood. It has a pattern that is predictable in some cases and, depending upon some circumstances, in other cases is not. The underlying foundation of the investigations has revealed that there are similar formations, structures, and dynamics that are in my opinion not random but carefully orientated to enable us to live. The electron orbits the nucleus of an atom in much the same way that moons orbit larger planets with the sun in the center. It would seem that just as a multiple number of planetary and moon systems can form the larger entity of a solar system structure, multiple atomic elements can form the larger structure of the molecule, which will in turn eventually into a galaxy or, for our discussion, matter.

Matter in motion has consequences for us. We can live among it as long as the motion is not disruptive. Elements of matter, that is, oxygen and hydrogen, form great masses that move in a circular rotation, just as a galactic structure would, and this dynamic can have an impact on our lives and can be dramatically altered for the worse. A hurricane looks like a galaxy from space; a thunderstorm looks like a towering form of harmless fluff from the ground. The details are in the joint action of its individual molecules within.

Because our atmosphere has a defined structure it can be subject to laws of physical influence. The greater the density of molecules, the greater the heat generated internally by self-compaction under the force of gravity. Because each atom weighs some finite amount, the entire formation then has weight, which we now understand as pressure. Heat, i.e., temperature, pressure, and density are an influence on the individual components that in turn react to that influence in an expectant way. This eventually forms the behavior that defines predictabilities.

In summary, the atmosphere has been charted from all its interlocking components. Just as a road map, navigation chart, or assembly instructions presents you with the details of road signs, road material, or expected signposts along the way and the larger picture of its connecting cities or landmasses, we have done the same. We have built from what was no knowledge of the invisible to an understanding of the atmospheric components that cannot be seen individually. We have started at the very beginning, when people first theorized there was indeed something in existence, but not what or why or how, and progressed to the inventions that enable us to see a clearer picture from the vastness of space and, communicate that observation to all on our planet below.

When we take a breath and look outward and upward, regardless of sun, cirrus, cumulus, or nimbus, an understanding of what had to occur to produce that observation and what that observation may mean in the next twelve to forty-eight hours will no doubt produce a satisfaction within. I wish to thank you for the charting of an atmospheric environment with me.

References

Ahrens, Donald C. *Meteorology Today*. Los Angeles: West Publishing Co. 1994.

Allaby, Michael. *A Chronology of Weather*. New York: Facts on File, 1998.

Amos, H.D., Lang, A.G.P. *These Were The Greeks*. Pennsylvania: Dufour Editions, 1996.

Army Times, The. *A History of the U.S. Signal Corps*. New York: G.P. Putnam's Sons, 1961.

AT&T. *Principles of Electricity*. USA: AT&T, 1961.

Barns, Jonathan. *The Presocratic Philosophers*. New York: Routledge & Kegan Press, 2000.

Barry, Roger G. and Chorley. Richard J. *Atmosphere, Weather & Climate*. New York: Routledge, 1998.

Bates, Charles C. and Fuller, John F. *America's Weather Warriors*. College Station: Texas A & M University Press, 1986.

Beauchamp, Ken. *History of Telegraphy*. London: Institution of Electrical Engineers, 2001.

Bohn, David. *Quantum Theory*. New York: Dover Publications, Inc., 1979.

Bolt, Bruce A. *Earthquakes and Geological Discovery*. New York: Scientific American Library, 1993.

Born, Max. *Atomic Physics*. New York: Dover Publications, Inc., 1969.

Branson, Lane K. *Introduction to Electronics*. New Jersey: Prentice-Hall, 1967.

Brown, Lloyd A. *The Story of Maps*. New York: Dover Publications, Inc., 1977.

Burr, Elizabeth. *The Chiron Dictionary of GREEK & ROMAN MYTHOLOGY*. Illinois: Chiron Publications, 1994.

Chabaud. *Weather: Drama of the Heavens*. 1994.

Coe, Lewis. *The Telegraph*. North Carolina: McFarland & Co., 1993.

Collier, Christopher G. *Applications of Weather Radar Systems*. England: Praxis Publishing, 1996.

Cottrell, T.L. *The Strengths of Chemical Bonds*. London: Buttersworth Scientific Publications, 1954.

Crowther, Arnold James. *Ions, Electrons, and Ionizing Radiation*. New York: Longmans, Green and Co., 1924.

Danielson, Eric W; Levin, James; Abrams, Elliot. *Meteorology*. New York: McGraw-Hill, 1998.

Davis, William Morris. *Elementary Meteorology*. Boston: Ginn & Company Publishers, 1894.

Dickenson, Oliver. *The Aegean Bronze Age*. United Kingdom: Cambridge University Press, 1994.

Doviak, Richard J. and Zrnic, Dusan S. *Doppler Radar and Weather Observations*. San Diego: Academic Press, 1993.

Dutton, John A. *Dynamics of Atmospheric Motion*. New York: Dover Publications, 1976.

Emanuel, Kerry A. *Atmospheric Convection*. New York: Oxford University Press, 1994.

Flemming, Roger James. *Meteorology in America, 1800–1870*. Baltimore: The Johns Hopkins University Press, 1990.

Gideons, *The Holy Bible*. Nashville: National Publishing Company, 1978.

Gillespie, Ronald J. and Popelier, Paul L.A. *Chemical Bonding and Molecular Geometry*. New York: Oxford University Press, 2001.

Gordon, Adrian, Grace, Warwick, Schwerdtfeger, Peter and Byron-Scott, Roland. *Dynamic Meteorology*. London: Hodder Headline Group and New York: John E. Wiley & Sons, 1998.

Gray, Harry B. *Chemical Bonds: An Introduction to Atomic and Molecular Structure*. Sausalito: University Science Books, 1994.

Guthrie, W.K.C. *The Greek Philosophers*. New York: Harper Torchbooks, 1950.

Hamblyn, Richard. *The Invention of Clouds*. USA: Farrar, Straus and Giroux, 2001.

Hamilton, Michael S. *Of The: Earth, Spheres and Consequences*. New York: Vantage Press, 2001.

Hargreaves, J.K. *The Solar-Terrestrial Environment: An Introduction to Geospace—the Science of the Terrestrial Upper Atmosphere, Ionosphere, and Magnetosphere*. New York: Cambridge University Press, 1995.

Haywood, John Ph.D. *Atlas of World History*. Spain: MetroBooks, 1977.

Herzberg, Gerhard. *Atomic Spectra and Atomic Structure*. New York: Dover Publications, 1944.

Hey, Tony and Walters, Patrick. *The Quantum Universe*. United Kingdom: The Cambridge University Press, 1997.

Huffman, Robert E. *Atmospheric Ultraviolet Remote Sensing*. Boston: Academic Press, Inc., 1992.

Jacob, Daniel J. *Introduction to Atmospheric Chemistry*. New Jersey: Princeton University Press, 1999.

Joos, Georg and Freeman, Ira M. *Theoretical Physics*. New York: Dover Publications, Inc., 1986.

Kearey, Philip, Brooks, Michael. *Introduction to Geophysical Exploration*. London: Blackwell Science Ltd., 1991.

Kidder, Stanley Q. Vondar and Harr, Thomas H. Vonder. *Satellite Meteorology an Introduction*. New York: Academic Press, 1995.
Kirk, G.S., Raven, J.E. and Schofield, M. *The Presocratic Philosophers*. United Kingdom: The Cambridge University Press, 1999.
Lang, Kennith R. and Gingerich, Owen. *A Sourcebook in Astronomy and Astrophysics*. England: Harvard University Press, 1979.
Laskin, David. *Braving the Elements*. New York: Doubleday, 1996.
Lemon, Harvey Brace, Ph.D. *From Galileo to Cosmic Rays*. Chicago: The University of Chicago Press, 1934.
Leveque, Pierre. *The Birth of Greece*. New York: Harry N. Abrams, Inc., 1994.
Lloyd, G.E.R. *Early Greek Science: Thales to Aristotle*. New York: W.W. Norton & Co., 1970.
Lutgens, Frederick K. and Tarbuck, Edward J. *The Atmosphere*. New Jersey: Prentice Hall, 2001.
McDonnel, John J. *The Concept of the Atom from Democritus to John Dalton*. United Kingdom: The Edwin Mellen Press, 1991.
Merril, Ronald T.; McElhinny, Michael W.; Mcfadden, Phillip L. *The Magnetic Field of The Earth*. San Diego: Academic Press, 1998.
Monmomier, Mark. *Air Apparent*. Chicago: University of Chicago Press, 1999.
Nardo, Don. *Greek and Roman Science*. San Diego: Lucent Books, Inc., 1998.
Pomeroy, Sarah B, Burstein; Stanley M., Donlan; Walter and Roberts and Jennifer Tolbert. *Ancient Greece*. New York: Oxford University Press, 1999.
Popper, Karl R. *The World of Parmenides*. London: Routledge, 1998.
Ramsey, William L. Phillips; Watenpaugh, Clifford R. and Frank M. *Modern Earth Science*. New York: Holt, Rinehart and Winston, 1965.
Sauvageot, Henri. *Radar Meteorology*. Boston: Artech House, Inc., 1992.
Seinfield, John H.; Pandis, Spyros N. *Atmospheric Chemistry and Physics*. New York: John E. Wiley & Sons, 1998.
Sill, William B.; Hoss, Norman. *Encyclopedia of the Sciences*. Baltimore: Popular Science Publishing Co. Inc., 1963.
Smith, Phyllis. *Weather Pioneers*. USA: Swallow Press/Ohio University Press, 1993.
Souyoudzoglou-Haywood, Christina. *The Ionian Islands*. Liverpool: Liverpool University Press, 1999.
Sullivan. J.W.N. *Atoms and Electrons*. New York: George H. Doran Co., 1924.
Tannoudji, Claude Cohen, Roc, Jacques Dupont, and Gryberg, Gilgert. *Atom-Photon Interactions*. New York: John E. Wiley & Sons, 1992.
Todd, Lewis Paul. *The Rise of the American Nation*. New York: Harcourt Brace Jonanovich, 1982.
Turcotte, Donald L., Schubert, Gerald. *Geodynamics*. New York: John E. Wiley & Sons, 1982.

Vasquez, Tim. *Weather Forecasting Handbook*. Texas: Weather Graphics Technologies, 2001.
Watters, Thomas R. *Planets*. New York: Macmillan, 1995.
Watts, Alan. *Instant Wind Forecasting*. New York: Sheridan House Publishing, 1988.
Watts, Alan. *The Weather Handbook*. Dobbs Ferry, NY: Sheridan House, Inc., 1999.
Williams, Jack. *The Weather Book*. New York: Vintage Books, 1997.
Williams, James Thaxter. *The History of Weather*. New York: Nova Science Publishers, 1999.
Zienert, Karen. *The Persian Empire*. New York: Benchmark Books, 1997.